Scientific Programming and Data Visualization in Mathematical Modeling

## Also of interest

*Differential Equations.*
*Solving Ordinary and Partial Differential Equations with Mathematica®*
Marian Mureşan, 2024
ISBN 978-3-11-141109-5, e-ISBN (PDF) 978-3-11-141139-2,
e-ISBN (EPUB) 978-3-11-141204-7

*Scientific Computing.*
*For Scientists and Engineers*
Timo Heister, Leo G. Rebholz, 2023
ISBN 978-3-11-099961-7, e-ISBN (PDF) 978-3-11-098845-1,
e-ISBN (EPUB) 978-3-11-098875-8

*Elementary Linear Algebra with Applications.*
*MATLAB®, Mathematica® and Maplesoft™*
George Nakos, 2024
ISBN 978-3-11-133179-9, e-ISBN (PDF) 978-3-11-133185-0,
e-ISBN (EPUB) 978-3-11-133195-9

*Machine Learning with Python.*
*Neural Networks, Algorithms, Deep Learning*
Sunil Karforma, Subhajit Adhikari, Indrajit De, Siddhartha Bhattacharyya, Jan Platos, 2026
ISBN 978-3-11-914470-4, e-ISBN (PDF) 978-3-11-221712-2,
e-ISBN (EPUB) 978-3-11-221739-9

# Scientific Programming and Data Visualization in Mathematical Modeling

With MATLAB® and Python

*By*

Mamta Kapoor, Geeta Arora

DE GRUYTER

**Mathematics Subject Classification 2020**
Primary 65-01; Secondary 00A71, 65Dxx, 68N15, 68U05, 65-04.

**Authors**
Dr Mamta Kapoor
Marwadi University Research Center
Department of Mathematics, Faculty of Engineering & Technology
Marwadi University, Rajkot, 360003 Gujarat
India
mamtakapoor.78@yahoo.com

Dr Geeta Arora
Department of Mathematics
Lovely Professional University
Phagwara 144411, Punjab
India
geetadma@gmail.com

ISBN 978-3-11-914699-9
ISBN 978-3-11-220775-8 (PDF)
ISBN 978-3-11-220782-6 (E-PUB)
DOI https://doi.org/10.1515/9783112207758

Library of Congress Control Number: 2026952958

Bibliographic information published by the Deutsche Nationalbibliothek
The Deutsche Nationalbibliothek lists this publication in the Deutsche Nationalbibliografie; detailed bibliographic data are available on the Internet at https://dnb.dnb.de.



De Gruyter and Walter de Gruyter GmbH are part of De Gruyter Brill.
www.degruyterbrill.com

Questions about General Product Safety Regulation: productsafety@degruyterbrill.com

Cover illustration: Monsitj/iStock/Getty Images Plus

# Contents

# Preface

Scientific programming has become an essential skill for students, researchers, engineers, data analysts, and professionals working in mathematics, science, engineering, and computational research. Modern mathematical modeling is no longer limited to theoretical formulation; it requires computational implementation, numerical experimentation, graphical interpretation, and data-driven analysis. The ability to translate mathematical ideas into efficient programs and meaningful visual outputs is therefore a fundamental requirement in contemporary scientific work.

This book has been written with the objective of providing a practical and comparative introduction to scientific computation using two widely used programming environments: MATLAB® and Python. MATLAB® is well known for its matrix-based numerical computing environment, integrated toolboxes, and strong applications in engineering, applied mathematics, and scientific computation. Python, on the other hand, is an open-source, flexible, and powerful programming language supported by libraries such as NumPy, Pandas, Matplotlib, and other scientific computing tools. Together, these two platforms offer a strong foundation for numerical computation, data handling, visualization, and mathematical modeling.

The main purpose of this book is to help readers understand not only the syntax of MATLAB® and Python but also their practical role in solving mathematical and scientific problems. The book begins with the fundamental programming environments, basic commands, variables, data types, arithmetic operations, scripts, functions, and data structures. It then gradually develops the reader's ability to work with vectors, matrices, arrays, conditional statements, loops, numerical computations, data analysis, visualization techniques, and symbolic computation. Wherever possible, the same concepts are explained in both MATLAB® and Python so that readers can compare their syntax, computational logic, strengths, and limitations.

The book is organized in a progressive manner. The introductory chapter presents the basic environments of MATLAB® and Python, including setup procedures, programming interfaces, basic syntax, variables, functions, and fundamental operations. The following chapters discuss MATLAB® fundamentals with special emphasis on vectors, arrays, and scientific calculations, followed by Python fundamentals involving lists, arrays, numerical libraries, and scientific computations. Subsequent chapters develop important programming concepts such as matrix operations, advanced array handling, conditional statements, loops, and user-defined functions. Further chapters focus on plotting, graphical representation, data visualization, computational mathematics, statistical analysis, numerical methods, and symbolic computation. The final part of the book includes

 | HTTPS://DOI.ORG/10.1515/9783112207758-203

computational practice questions and project-based problems that allow readers to apply MATLAB® and Python to mathematical modeling, numerical solutions, error analysis, and comparative scientific computation.

A special feature of this book is its emphasis on practice-based learning. Each topic is supported by examples, explanations, computational exercises, short-answer questions, and project-based problems. These features are intended to help readers move from passive reading to active implementation. The examples are designed in a simple and progressive manner so that beginners can follow them comfortably, while advanced learners can extend them to more complex mathematical models, simulations, and research problems.

Data visualization has also been given significant importance in this book. In mathematical modeling and scientific computing, visualization is not merely a method of presenting results; it is a powerful tool for understanding patterns, interpreting model behavior, comparing solutions, identifying errors, and communicating findings effectively. Through MATLAB® plotting tools and Python visualization libraries, readers are introduced to practical ways of representing numerical results graphically and meaningfully.

This book is suitable for undergraduate and postgraduate students, research scholars, faculty members, engineers, applied mathematicians, and data science learners who wish to strengthen their computational skills. It may also serve as a useful reference for those who are transitioning from MATLAB® to Python or from Python to MATLAB®, especially in the context of mathematical modeling, scientific computation, and data visualization.

The readers are encouraged to run the programs, modify the examples, compare outputs, and explore alternative approaches. Programming is best learned through consistent practice, and mathematical modeling becomes meaningful when theory, computation, visualization, and interpretation are combined. It is hoped that this book will help readers develop confidence in using MATLAB® and Python as effective tools for scientific programming, data visualization, and computational mathematical modeling.

CHAPTER 1

# Introduction to MATLAB and Python

A number of **tasks** that **need** to be performed in the study of data, or to **analyze** scientific **phenomena**, **require** computational **skills.** There are a number of **computational tools** that can be utilized to operate on data and **obtain** results. The two most **popular** tools quite often **opted for** by the **research community** are **MATLAB** and **Python.** While MATLAB offers an integrated environment to perform the calculation, Python is an open-source software emerging as a unbeatable tool to perform scientific calculations. MATLAB has the foundation on the concept of matrix and has application in solving the engineering problems.

The choice can be made between the two based on the requirement of the user as per the context in which the calculation is required. It also depends on the availability of the software, the licensing constraints, and the specific requirements of the computational task. This chapter explores the strengths of both the languages to solve the scientific problems efficiently starting from the basic commands and exploring for the implementation. This chapter provides a concise introduction to both environments and their basic programming constructs.

By the end of this introductory chapter, the user will be able to:
- Get equipped with the differences and strengths of MATLAB and Python
- Set up the programming environments for both the programming languages
- Learn the fundamental operations in both the languages
- Gain the confidence in handling the programming language efficiently

The following is the structure of the content presented in this chapter:
- Overview of MATLAB with its setup and environment details with basic concepts
- Set up the Jupyter IDE for Python and its basic commands to perform the basic operations

## Detailing MATLAB

MATLAB is a high-performance numerical computing programming language developed by Cleve Moler handled by MathWorks. It's a short form of the MATrix LABoratory and provides an interactive platform for numerical computation, algorithm development, and data handling. Due to its powerful toolboxes and interactive interface, it is a favorite choice of the researchers from the diverse computational background requirements.

 | HTTPS://DOI.ORG/10.1515/9783112207758-001

Where to get MATLAB from?

- The MATLAB can be downloaded and purchased on the MathWorks website or even can be worked on the online MATLAB creating an account.
- Or there is a substitute of Octave online in case you are a novice and just aim to look for its initial usage from the webpage: https://octave-online.net/
- There is an open-source software called as Scilab in which the users can practice and run the MATLAB code and programs but has some notable differences that be seen when some ambiguity appears. Download it from: https://www.scilab.org/download

Thus, in a nutshell for the programmer eager to learn and use the MATLAB, there is no barrier. So, let's start with the environment details that appear on the launch of the software.

## MATLAB environment

On the launch of the MATLAB, there are several key components of the window that appears. Figure 1.1 presents the window in its standard form.

**Command Window:** The primary interface of MATLAB where a user can execute commands, perform calculations, and interact with the MATLAB environment in real time is the command window. It can be referred as the console with a prompt waiting from the input and ready to execute the entered command. It is the interface that interacts with users mentioning the bugs, giving the output, and even providing the help to the topics.

**Workspace:** The dynamic area of the MATLAB interface to display the variables that are active in the memory is the workspace. User can navigate it to see the details about the variables and monitor their size, lengths, and the type from this small window. It displays the variables name their values and the data types.

There are commands also that can help displaying all the information of variables. For example: who and whos are the commands that when entered on the command window provide the necessary details of the variables.

**Current Folder:** This part of the MATLAB showcases the active area where the files are going to be saved. The current folder needs to be changed as per the folder where the files are appearing. The **Current Folder** shows every file, script, and dataset, which are currently in use. It provides easy access to move between folders using the top bar or the cd command.

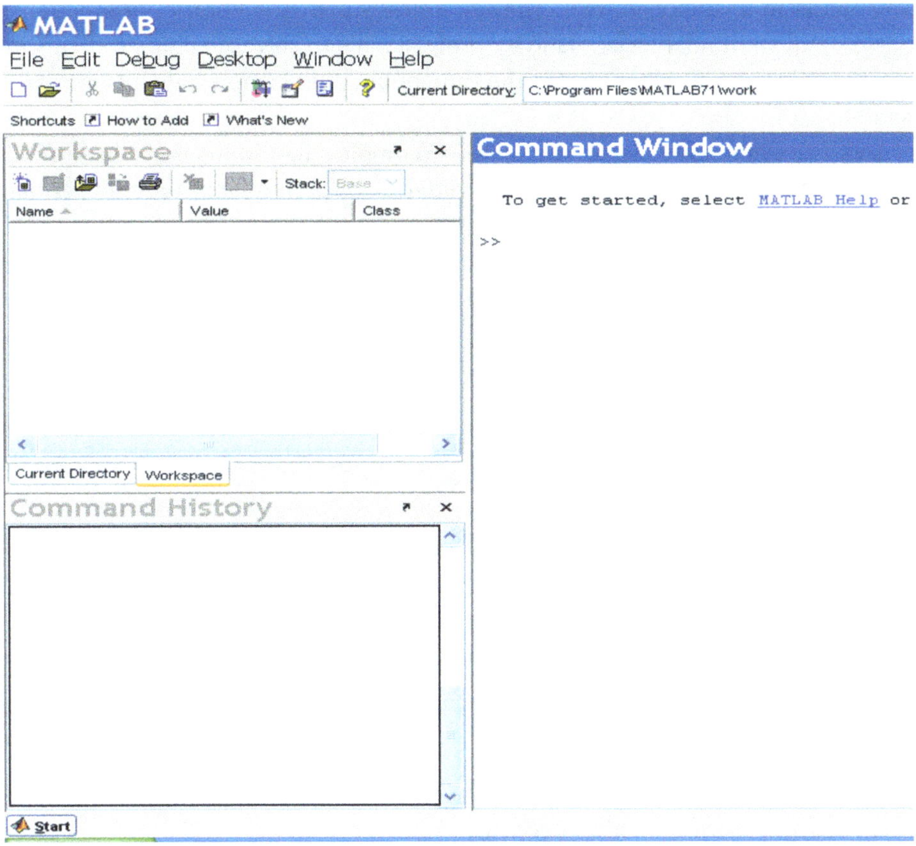


FIGURE 1.1 Layout of MATLAB interface.

**Editor:** The **MATLAB Editor** is the primary workspace for developing reusable code as a program. Unlike the **Command Window** which is best place to work with the single commands, the Editor provides a structured environment for building complex scripts and functions. The Editor is packed with features designed to improve code readability and reduce manual errors. It includes the features like **syntax highlighting** and **automatic indentation** make the logic easier to follow at a glance. It also includes tools such as sectioning of the code and **tabbed browsing** that allow user to navigate large projects. Furthermore, the real-time **error checking** highlights potential bugs and thus the Editor serves as a diagnostic lab. The **breakpoints** can be set to pause execution at specific lines, allowing to step through the logic and inspect variables in real-time.

**Toolboxes:** The power of the MATLAB environment, both in professional and academic settings, comes from its toolboxes. The basic MATLAB package is a good basic tool for general matrix-math and plotting. The toolboxes add specific capabilities in a variety of areas. They're basically premade libraries of functions, classes, and simulink blocks made by professionals to solve specific problems in specific industries. Toolboxes save the time of building complex algorithms from scratch.

Toolboxes eliminate the need to develop complex algorithms from the ground up. Instead, they provide a framework of vetted, high-performance tools for various disciplines, including specialized suites for **signal processing**, **control systems**, and **image processing**; advanced environments for **machine learning**, **deep learning**, and **statistics**; and sophisticated solvers for linear, quadratic, and mixed-integer programming.

## Basic syntax and operations

### Comments

MATLAB uses the percent symbol (%) for comments. It also offers a unique "Block Comment" feature and "Sections" for organizing code within the Editor.

- **Single-line comment:** Everything following the % on that line is ignored by the interpreter.
  **Example:**

```
% This is a single-line comment in MATLAB
x = 10; % You can also comment at the end of a line
```

- **Block (multiline) comment:** Uses the %{ and %} delimiters. This is useful for long explanations or temporarily disabling large chunks of code.
  **Example:**

```
%{
   This is a block comment.
   It can span multiple lines
   without needing a % on every line.
%}
```

- **Sections:** Two percent signs %% create a code section (or "cell"), which can be executed independently in the MATLAB Editor.

**Example:**

```
%% Section 1: Data Analysis
```

### Variables and data types

The rules that should be followed to define variables are given below:
- Must begin with a letter (e.g., data1 is valid, 1data is not)
- **Allowed characters:** Only letters, numbers, and underscores
- **No spaces or special characters:** Don't use hyphens, dots or symbols (e.g., my_var is valid, my-var is not)
- **Case sensitivity:** temp, Temp, and TEMP are three distinct variables
- **Reserved words:** Do not use the MATLAB keywords (e.g., if, for, while) as a variable name

Type is keyword to see the list in full.

MATLAB reserved words are The MATLAB reserved words include break, case, catch, classdef, continue, else, elseif, end, for, function, global, if, otherwise, parfor, persistent, return, spmd, switch, try, and while.

Basic control flow and structural commands define how programs are written and executed. These include conditional branching (if, else, elseif, switch, case, otherwise), looping (for, while, parfor, spmd), error handling (try, catch), and program organization (function, classdef, global, persistent). The break and continue control the execution of loops, return indicates the end of functions, and end signifies the end of blocks. These words together form the core of MATLAB programming logic.

- **Length limit:** Names must be shorter and meaningful (can have maximum 63 characters).

Variables in a programming language can be written in different naming styles. Table 1.1 showcased the common naming styles along with the explanation and description.

TABLE 1.1 Different ways to name variables.

| Style | Example | Description |
|---|---|---|
| **Camel case** | studentName | First word lowercase, next words capitalized |
| **Snake case** | student_name | Words separated by underscore |
| **Pascal case** | StudentName | Each word capitalized |

### Arithmetic operations

Arithmetic operations form the foundation of numerical computation in MATLAB. These operations are carried out using familiar mathematical operators as follows:

Addition (+), subtraction (–), multiplication (*), and division (/ for right division, \ for left division).

Beyond these, MATLAB provides element-wise operators as .* for multiplication, ./ and .\ for division, and .^ for exponentiation. These operators allow operations to be applied directly to corresponding elements of arrays.

Element-wise operations are particularly required while working with datasets or simulations that require computations across individual entries. MATLAB adheres to conventional operator precedence rules, ensuring predictable evaluation order, while parentheses can be used to explicitly control the sequence of operations.

The examples that can be tried on the console for the insight with the mathematical operators are shown in Table 1.2.

TABLE 1.2 Examples of expressions and their output.

| Expression | MATLAB syntax | Output |
|---|---|---|
| 2 - 5 + 9 | 2 - 5 + 9 | 6 |
| $8^2$ | 8^2 | 64 |
| 2 / 5 | 2 / 5 | 0.4000 |
| 2 \ 5 | 2 \ 5 | 2.5000 |
| 6 x 7 | 6 * 7 | 42 |
| 6^{(3+4)} | 6^(3+4) | 279936 |

### Functions and scripts

In the MATLAB environment, both scripts and functions serve as program files containing sequences of commands. However, they operate under different logic regarding data handling, variable scope, and modularity.

### MATLAB functions

Functions are the structured modular program files which are designed to process the inputs and return specific results. The functions operate within a **local workspace** and ensure that variables defined inside a function do not overwrite or inter-

fere with data in the base workspace, unless those values are explicitly returned as outputs. This makes functions a preferable choice for repetitive tasks or complex calculations requiring different datasets. Following are the key characteristics of function that make them unique as per the input and output requirements:
- **Definition:** Initiated with the function keyword
- **Data Flow:** In particular, taking input arguments and returning output values
- **Scope:** Isolates the variable workspace to guarantee data integrity
- **Modularity:** Supports reusable and structured programming

The general syntax of the function in MATLAB is as follows:

```
function [ output_args ] = Untitled( input_args )
%UNTITLED Summary of this function goes here
% Detailed explanation goes here


end
```

Just like a new file can be opened to create program, a function can be opened using the menu and saved with the same name as of that of the defined function.

### Scripts

Scripts are .m files that store a simple, linear sequence of MATLAB commands. Unlike functions, scripts operate directly within the base workspace. That is, any variable you create or change while the script is running will still be available in your command window memory after the script is finished. Scripts are most useful when you are automating a sequence of calculations, creating visualizations, or running simulations and the data is already loaded or defined in the file.

**Scripts features:**
- **No explicit I/O:** They don't have any formal input arguments or any specific output values.
- **Shared workspace:** The global MATLAB workspace is shared and existing variables can easily be overwritten.

The major differences in script and function files are presented in Table 1.3, which can be selected as per the requirements.

TABLE 1.3 Different features of the script files.

| Feature | Scripts | Functions |
|---|---|---|
| **Workspace** | Shares the base workspace | Uses a private local workspace |
| **Arguments** | Does not accept inputs | Accepts input arguments |
| **Outputs** | Saves variables to the workspace | Returns specific output variables |
| **Best For** | Quick automation and analysis | Modular tasks and reusable logic |

## Detailing Python

Python is a versatile, high-level programming language designed with a primary emphasis on code readability and simplicity. As an interpreted language, it executes code line-by-line, which eliminates the need for a separate compilation step. This allows for **rapid prototyping** and immediate feedback, making it a favorite for both beginners and professional developers.

Python is often a preferred first choice for novice coders due to its **syntax**, which matches with the syntax used in daily computational task. Unlike languages like C++ or Java, Python uses indentation to define code blocks instead of complex curly braces {} or semicolons ;. This allows developers to write clean, organized code.

One of the primary reasons for choosing Python is its extensive collection of libraries. Python has a standard library that allows programmer to write a program in a predefined form that includes working with the data extraction, manipulation, as well as analysis. Furthermore, Python also provides several predefined modules designed to perform specific tasks. Following are the specialized libraries designed for high-level data tasks:

- **Pandas:** Used for data manipulation and analysis. It provides data structures like DataFrames, which allow users to easily clean, filter, and reshape structured data.
- **NumPy:** Short for "Numerical Python," it is the foundation for scientific computing. It enables efficient operations on large, multidimensional arrays and matrices, providing a massive collection of high-level mathematical functions.
- **Matplotlib/Seaborn:** These are the go-to libraries for data visualization, allowing user to turn complex datasets into intuitive charts, histograms, and heatmaps.
- **Scikit-learn:** This is a machine learning library, providing simple and efficient tools for predictive data analysis, such as regression, classification, and clustering.

The key features of Python that make it especially appealing to developers includes its interactive workflow where small snippets of code can be tested instantly, speeding up experimentation and debugging. Its **high-level nature** provide solution to the complex tasks by letting developers focus on solving problems rather than handling low-level details. The **cross-platform capability** of Python ensures that code written on Windows can run seamlessly on macOS or Linux without modification. Also, the **vast community** of Python provides endless support through tutorials, documentation, and forums like Stack Overflow, Github thus making it easier for developers to learn, troubleshoot, and grow their skills.

## Python environment

Python is a versatile, high-level programming language designed with a focus on readability and simplicity. It is an interpreted language, executing the code sequentially, line by line. A distinct compilation phase is not required. This facilitates a quick interaction environment for prototyping and prompt feedback, making it a preferred choice for both novices and experienced developers.

Python is frequently the preferred initial choice for novice programmers due to its syntax similar to other programming language.

Python employs indentation to delineate code blocks, in contrast to the intricate use of curly braces {} or semicolons ; found in C++ or Java. Developers can produce clean and organized code utilizing it.

The extensive array of libraries provided by Python is a primary justification for its selection. Python possesses a standard library that enables programmers to compose programs in a predefined manner for the data extraction, manipulation, and analysis. Python possesses numerous predefined modules. These are utilized to execute particular tasks. Some of the specialized libraries for advanced data operations are as follows:

- **Pandas:** It is used for data manipulation and analysis. It provides data structures like DataFrames that facilitate the cleaning, filtering, and reshaping of structured data.
- **NumPy:** It denotes "Numerical Python." It constitutes the foundation of scientific computing. It facilitates efficient operations on extensive multidimensional arrays and matrices. It additionally provides an extensive array of advanced mathematical functions.
- **Matplotlib/Seaborn:** These are the principal libraries employed for data visualization, enabling users to transform intricate datasets into comprehensible charts, histograms, and heatmaps.

- **Scikit-learn:** This is a machine learning library that provides straightforward and effective tools for data analysis and predictive modeling, encompassing regression, classification, and clustering.

This is particularly attractive to developers because of the interactive workflow provided by Python.

Python is cross-platform, allowing code written on Windows to execute immediately on macOS or Linux without modifications. Moreover, the extensive Python community provides boundless assistance via tutorials, documentation, and platforms such as Stack Overflow and GitHub, facilitating developers in learning, debugging, and refining their skills.

## Python ecosystem

To commence programming in Python, one must first establish a functional workspace. This encompasses the installation of the Python interpreter and the selection of an environment conducive to coding, utilizing Python's available libraries and packages.

Obtain the most recent stable version of Python for your operating system (Windows, macOS, Linux) from the official website python.org.

The standard installation package includes:

- The Python interpreter: The mechanism that executes your code
- The standard library: A compilation of intrinsic modules that executes conventional functions
- IDLE: A user-friendly integrated development environment for novices

### Crucial notice for Windows users

During installation, verify that the option "Add Python to PATH" is selected. Python can be executed directly from the command line or terminal.

### Alternative Anaconda distribution

A significant number of users, particularly in data science and academia, favor the installation of Anaconda. This widely used distribution integrates Python with essential scientific libraries, including NumPy, Pandas, and Matplotlib, along with tools such as Jupyter Notebook. It enables the construction of intricate machine learning and analytical workflows.

### Choosing an integrated development environment or code editor

Selecting the appropriate integrated development environment (IDE) or code editor is crucial for enhancing productivity, debugging efficacy, and coding convenience. The most prevalent environments are:

- **IDLE:** Python's default lightweight IDE. It is simple and features an integrated shell. Ideal for novices crafting concise scripts.
- **PyCharm:** A professional-grade integrated development environment created by JetBrains. Advanced code completion, version control, and optimized for extensive software projects.
- **Spyder:** Tailored for scientific computation. It features an interface akin to MATLAB, complete with a variable explorer and plot viewer. It is frequently incorporated in the Anaconda distribution.
- **Thonny:** An integrated development environment tailored for educational use, featuring an intuitive user interface and incremental debugging capabilities.

**The following are general-purpose editors:**
- **Visual Studio Code (VS Code):** A lightweight, extensible code editor developed by Microsoft. The Python extension serves as a robust tool for both novice and experienced developers.
- **Jupyter Notebook:** A web-based interactive environment. It enables the integration of live code with narrative text and visualizations, representing the pinnacle of data analysis and reproducible research.
- **Sublime Text/Atom:** Efficient and nimble editors, suitable for rapid modifications and lightweight scripting.

The Anaconda Distribution is the most highly recommended initial resource for novice programmers learning Python. Anaconda is a comprehensive package that installs the Python programming language, essential data science libraries (such as NumPy and Pandas), and intuitive coding environments.

Here are the details that one must adhere to:

1. **Download and install**

   Obtain the installer from anaconda.com. Unless you possess advanced expertise, it is recommended to select the default options during the installation procedure. Upon completion of the installation, the user possesses a comprehensive suite of tools available for immediate use, without the necessity for additional internet downloads.

2. **Initiating Jupyter Notebook**

   Launch the Jupyter Notebook application to commence coding. There are two methods to accomplish this.

   **The Visual Technique:** Access Anaconda Navigator from your applications menu as shown in Figure 1.2 from here, and select the "Launch" button for Jupyter Notebook.

   **The Command Way:** Open the terminal (or Anaconda Prompt on Windows), type jupyter notebook, and hit Enter.

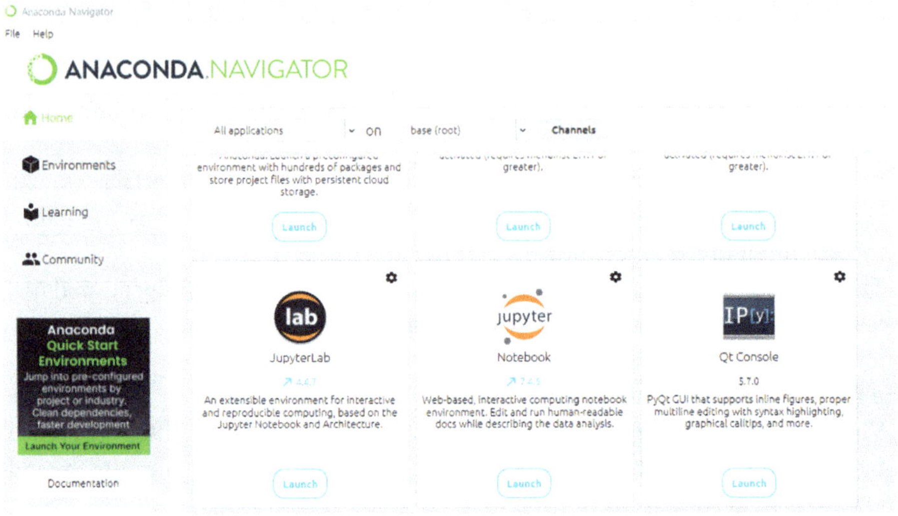


FIGURE 1.2 Window view of Anaconda Navigator showcasing inbuilt modules.

3. **Using the web interface**
   One unique feature of Jupyter is that it does not open in a traditional software window. Instead, it opens within your default web browser (such as Google Chrome, Firefox, or Microsoft Edge):
   - **Local server:** Even though it appears in a browser, it is running locally on the system and hence requires no internet connection to run the code.
   - **File dashboard:** The first thing that appears on launch is a file directory of the system. User can navigate to the "Documents" or a specific project folder before creating a new notebook file. To create a new file, navigate to File-New. User can select here the Notebook to create a new notebook to work with the commands as shown in Figure 1.3.

The output that opens in the new tab after choosing New-Notebook is shown in Figure 1.4.

This is the workspace where Python commands can be written and executed. A cell can be executed using the keyboard shortcut Shift + Enter.

## Working in Jupyter Notebook

Jupyter Notebook supports two types of cells:

1. **Code cells**
2. **Markdown cells**

Python is a high-level programming language widely used for scientific computing and data analysis. Markdown allows notebooks to function like **interactive docu-**

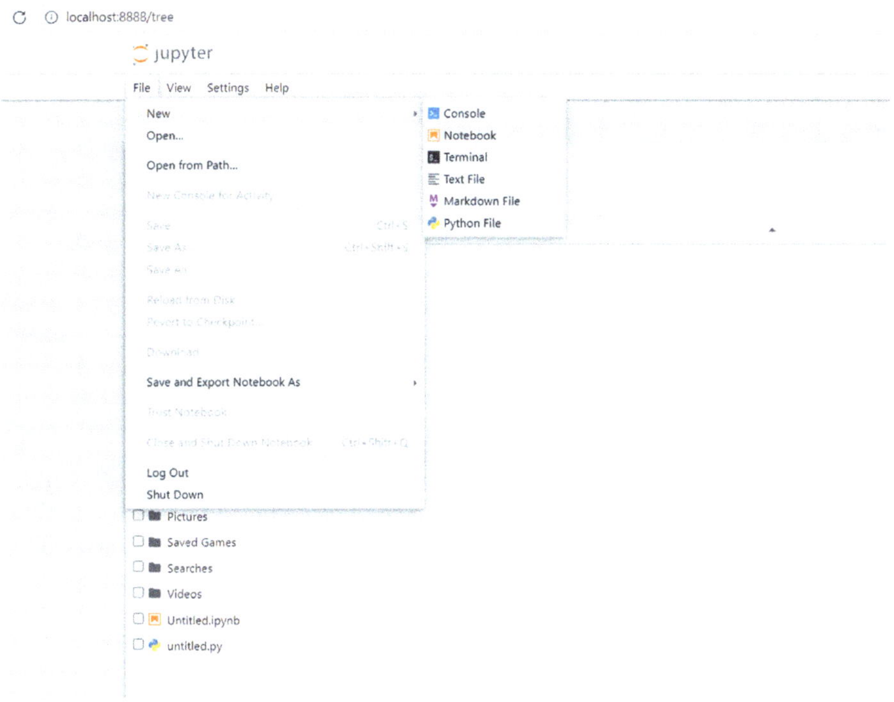


FIGURE 1.3 Interface of Jupyter to create a new file.

FIGURE 1.4 Cell of the Jupyter notebook

**ments** combining **text, code, and output.** This makes Jupyter Notebook an excellent tool for **documentation, teaching, and reporting results.**

### Converting a cell to Markdown

To convert a cell to Markdown:

1. Select the cell
2. Press **Esc**
3. Press **M**

Alternatively, select **Markdown** from the dropdown menu. This is how the numbering on the left will disappear and make it work as a comment (see Figure 1.5).

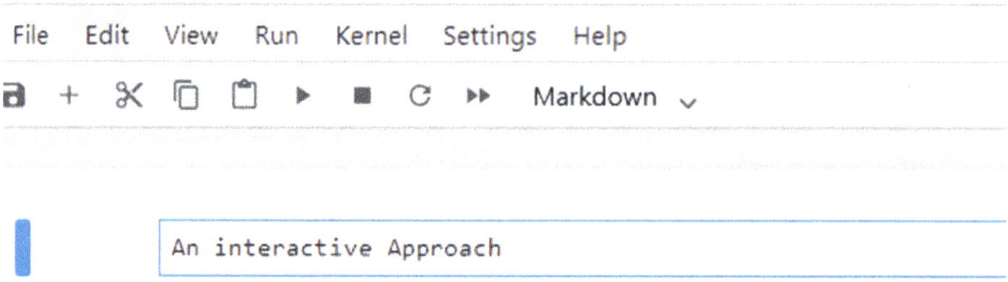


FIGURE 1.5 Showcasing the Markdown tab to provide the comment.

## Checking the Python version

It is often useful to check the version of Python currently being used. This can be done using the **sys module**, which provides system-related functions and then printing the version with the command **print(sys.version)** (see Table 1.4). This command is useful when working across different systems or verifying compatibility with libraries.

TABLE 1.4 A visual to understand the layout to validate the version of Python in use.

```
import sys
print(sys.version)
3.13.9 | packaged by Anaconda, Inc. | (main, Oct 21 2025, 19:09:58) [MSC v.1929 64 bit
(AMD64)]
```

## Print function in Python

The print() function in Python is used to display information on the screen. It is one of the most frequently used commands when learning programming because it allows the user to see the results of computations, messages, or values stored in variables.

The basic syntax of the print function is:

**print(object)**

Here, object can be text, numbers, variables, or expressions whose value needs to be displayed.

### Printing message

print("Hello World")

**Output**
Hello World
Text can be written using either single quotes (' ') or double quotes (" ").
print('Python Programming')
print("Python Programming")
Both statements produce the same output.

### Printing variables

The print() function can also display the value or text stored in a variable as shown in Table 1.5.

TABLE 1.5 Various commands involving print option along with the output.

| |
|---|
| **x = 10** |
| **print(x)** |
| **10** |
| **name = "Anil"** |
| **print(name)** |
| **Anil** |

### Printing multiple variables in Python

The print() function allows multiple values to be displayed in a single command. The values are separated using commas. For example, see Table 1.6.

General syntax:

**print(value1, value2, value3)**

TABLE 1.6 Presentation of the print command taking more than one input.

| |
|---|
| **x = 10** |
| **y = 20** |
| **z = 30** |
| **print(x, y, z)** |
| **10 20 30** |

### Printing text and variables together

The text values and variables can be printed together using the print command (see Table 1.7). It is very useful when displaying results in a readable form.

TABLE 1.7 Presentation of the input and output for printing the variables with content.

```
name = "Ravi"
age = 21

print("Name:", name, "Age:", age)
Name: Ravi Age: 21
```

Variables can also be joined using the + operator, but this works only for strings (Table 1.8). However, the comma method is generally preferred because it automatically handles different data types.

TABLE 1.8 Presentation of the layout of getting the output using print with text and the variable.

```
Name = “Ravi”
print("Hello " + name)
Hello Ravi
age = 20
print("Age:", age)
Age: 20
```

## Basic syntax and operations

### Comments

Python uses the hash symbol (#) for standard comments. While Python does not have a formal "block comment" symbol like MATLAB's %{ %}, it uses multiline strings for documentation:

– **Single-line comment:** Uses the # symbol. This is the standard way to document code.
  **Example:**

```
# This is a single-line comment in Python
x = 10 # This is an inline comment
```

– **Multiline strings (Docstrings):**
In Python, triple quotation marks (""" """ or ''' ''') are used for creating multiline string literals that allows the text to span multiple lines without using newline characters.

When a multiline string is placed as the first statement inside a module, function, class, or method, it is treated as a docstring (documentation string). A docstring provides structured documentation that describes the purpose and usage of the associated code element.

**Docstrings are commonly used to explain:**
- Regarding the purpose of a module, function, or class
- For input parameters and expected data types
- To return values produced by the function
- For additional notes or usage examples

The details of examples are shown in Table 1.9.

TABLE 1.9 Presentation of examples of comments in different forms.

```
"""
This is a multi-line string used as a module docstring.
It describes the purpose of the Python script.
"""

def add(a, b):
    """
    Returns the sum of two numbers.

    Parameters:
        a (int or float): First number
        b (int or float): Second number

    Returns:
        int or float: Sum of a and b
    """
    return a + b
```

## Variables and data types

In Python, **variables** are the symbolic names using which data can be referenced and manipulated. Conceptually, a variable is a named location in the computer's memory that stores a value. Unlike many other programming languages, a variable in Python is created when a value is assigned to it using the **assignment operator** (=).

Python is a **dynamically typed** language; thus, there is no need to explicitly declare the data type (such as "integer" or "string") when creating a variable (just like MATLAB). The Python interpreter automatically infers the type based on the value assigned to it. Furthermore, a variable can change its type during execution if it is reassigned to a different kind of value.

Data types define the nature of the data and what operations can be performed on it. Following are the built-in data types of Python:

- **Numeric types:** int (integers like 10), float (decimal numbers like 3.14), and complex
- **Sequence types:** str (text/strings), list (mutable ordered collections), and tuple (immutable ordered collections)
- **Mapping type:** dict (key-value pairs)
- **Set types:** set (unordered collections of unique items)
- **Boolean type:** bool (represents True or False)
- **None type:** Represented by the keyword None, used to signify the absence of a value

**Example of variable assignment:**
Python

```
x = 50          # x is an Integer
y = 2.181       # y is a Float
name = "Program" # name is a String
is_active = True # is_active is a Boolean
```

Since the variable naming convention of MATLAB is already discussed, Python also follows some similar rules such as case-sensitive and generally flexibility.

While MATLAB variables must start with a letter and can contain up to 63 characters, Python has slightly different technical boundaries as follows:

**Leading characters:** In MATLAB, a variable should start with an alphabet. However, in Python, a variable can be started with an **underscore (_)** and has a special meaning (often indicating "internal" or "private" variables). In MATLAB, underscores are typically used only within the name.

**Reserved keywords:** Python also has a specific list of reserved keywords that cannot be used as variable names (e.g., None, True, lambda, with, yield). User can check these in Python using import keyword, print(keyword.kwlist).

The way to write and the expected output are shown in Figure 1.6.

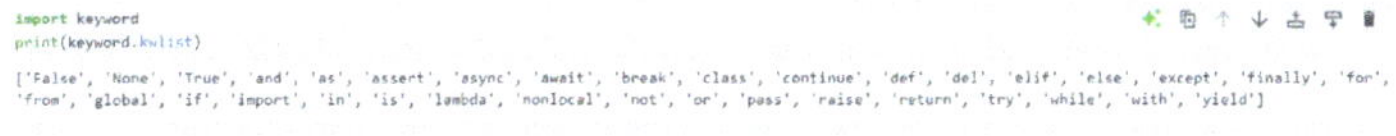

```
import keyword
print(keyword.kwlist)

['False', 'None', 'True', 'and', 'as', 'assert', 'async', 'await', 'break', 'class', 'continue', 'def', 'del', 'elif', 'else', 'except', 'finally', 'for',
'from', 'global', 'if', 'import', 'in', 'is', 'lambda', 'nonlocal', 'not', 'or', 'pass', 'raise', 'return', 'try', 'while', 'with', 'yield']
```

FIGURE 1.6 Showcasing the output of keywords of Python.

Regarding case sensitivity, both the languages are case-sensitive (Data and data are different). Table 1.10 presents a comparison of how variable naming can be differentiated in both languages.

TABLE 1.10 Discussion of the naming rule comparing MATLAB and Python.

| Feature | MATLAB rule | Python rule |
| --- | --- | --- |
| **Primary style** | Often camelCase | Strictly snake_case |
| **Start character** | Must be a letter | Letter or underscore (_) |
| **Length limit** | 63 characters (namelengthmax) | Theoretically unlimited |
| **Constants** | No strict naming rule | Use ALL_CAPS (e.g., PI = 3.14) |
| **Built-in protection** | Stronger protection/warnings | Easy to accidentally overwrite |

## Arithmetic operations

Arithmetic operations allow a user to perform mathematical calculations using a set of dedicated operators. Python supports standard almost all of the mathematical operations that can be applied to numeric data types like integers and floats.

Beyond standard addition and subtraction, Python provides some specialized operators as follows:

**floor division** (//): It returns the largest integer less than or equal to the result (discarding the decimal)

**modulus** operator (%): It returns the remainder of a division.

Python follows the standard mathematical rules of precedence – often referred to as **PEMDAS** (parentheses, exponents, multiplication/division, addition/subtraction) – to evaluate complex expressions.

Table 1.11 summarizes the most common arithmetic operators in comparison to what is discussed for the MATLAB. This summary is intended to provide a basic insight of the similarity the changes that need to be deal with.

TABLE 1.11 Discussion of the syntax to perform the mathematical operation in both MATLAB and Python.

| Mathematical expression | MATLAB syntax | Python syntax | Output |
|---|---|---|---|
| **2 - 5 + 9** | 2 - 5 + 9 | 2 - 5 + 9 | 6 |
| **$8^2$** | 8^2 | 8 ** 2 | 64 |
| **2 \ 5** | 2 / 5 | 2 / 5 | 0.4 |
| **5 \ 2 (Left Div.)** | 2 \ 5 | 5 / 2 | 2.5 |
| **6 x 7** | 6 * 7 | 6 * 7 | 42 |
| **6^{(3+4)}** | 6^(3+4) | 6 ** (3 + 4) | 279936 |
| **10 ÷3 (Integer)** | fix(10/3) | 10 // 3 | 3 |

## Compound assignment operators in Python

In Python, compound assignment operators combine an arithmetic operation with assignment. These operators allow a variable to be updated by performing an operation on its current value and assigning the result back to the same variable.

Instead of writing a long expression such as x = x + 5, Python allows the same operation to be written more concisely using a compound assignment operator:

x += 5

Thus, compound assignment operators help make the code **shorter, clearer, and easier to maintain.**

**General Syntax**

```
<Variable> <Operator>= <value>
```

This expression performs the arithmetic operation first and then assigns the result back to the variable. Almost all the operators can be compounded with the **assign-**

**ment operator.** The details of the operators and their meanings are provided in Table 1.12 with the examples in Table 1.13.

TABLE 1.12 Discussion of the operators along with the details.

| Operator | Meaning | Equivalent expression |
|---|---|---|
| += | Addition assignment | x = x + value |
| -= | Subtraction assignment | x = x - value |
| *= | Multiplication assignment | x = x * value |
| /= | Division assignment | x = x / value |
| //= | Floor division assignment | x = x // value |
| %= | Modulus assignment | x = x % value |
| **= | Exponent assignment | x = x ** value |

To understand the assignment, consider the examples in the following table:

TABLE 1.13 Examples of commands performing mathematical operations.

```
x = 10
x += 5
print(x)
15
x-=10
print(x)
5
x**=2
print(x)
3600
```

## Specific data structures: lists, tuples, and dictionaries

While MATLAB is built with the focus around the **Matrix** – to deal with the uniform numerical data, Python is a general-purpose language designed to handle diverse types of information. To manage this, Python introduces some specific data structure: **lists, tuples, and dictionaries.** These structures are essential that allow programmer to group together different data types (like mixing text and numbers) and organize complex data by labels rather than just row and column indices.

### 1. Lists: the dynamic container

A list in Python is similar to a one-dimensional MATLAB cell array. While a MATLAB vector is optimized for math, a Python list is optimized for storage and collection.

One can mix integers, strings, and even other lists within a single set of **square brackets.** Furthermore, lists are mutable, that can be grown, reduced, or changed without redefining the entire variable.

A list is best used when there is a collection of items that might change, such as a list of experimental trials or a series of names.

**Example:**

```
# A list of mixed data types
experiment_results = [10.5, "Success", 12.2, "Failure", 9.8]

# You can modify a list (Mutability)
experiment_results[1] = "In Progress"

# You can add items
experiment_results.append(11.4)

print(experiment_results)
# Output: [10.5, 'In Progress', 12.2, 'Failure', 9.8, 11.4]
```

### 2. Tuples: the "Locked" list

A **tuple** looks like a list but is defined with **parentheses ().** It can be considered as a "Read-Only" list. Once created, it cannot be changed.

Tuples are used for data that belongs together and should **never** be changed accidentally during the program, such as geographic coordinates or RGB color values.

**Example:**

```
# A tuple representing a GPS coordinate (Latitude, Longitude)
location = (40.7128, -74.0060)

# If you try to change it:
# location[0] = 41.0 --> This will cause a TypeError!
```

```
print(location[0])
```

# Output: 40.7128

3. **Dictionaries: mapping by label**

A Python **Dictionary** allows the user to access data using **Keys** (labels), which can be defined using **curly braces {}.**

**Example:**

```
# A dictionary representing a sensor's metadata
sensor_data = {
    "id": "SN-502",
    "type": "Thermometer",
    "reading": 22.5,
    "unit": "Celsius"
}

print(sensor_data["reading"])
# Accessing data by the label (Key)
# Output: 22.5
# Adding a new key-value pair
sensor_data["status"] = "Active"
```

### Functions

A function is a reusable block of code designed to perform a specific task. Functions help organize programs into smaller modules and avoid repetition. In Python, functions are defined using the **def** keyword.

**Syntax of a function**

```
def function_name(parameters):
statements
return value
```

Example of defining and calling a function without argument is presented in Table 1.14 and the function with arguments in Table 1.15.

TABLE 1.14 Defining and calling a function.

```
def greet():
    print("Welcome to Python Programming") # defining function
greet()                                    # Calling functions
Welcome to Python Programming              # output
```

Functions can also be defined that can accept inputs known as **parameters.**

### Example of a function with arguments

TABLE 1.15 Defining the function along with arguments followed by calling a function.

```
def add_numbers(a, b):
    result = a + b
    return result

add_numbers(5,3)
8
```

### NumPy for numerical computing

NumPy (numerical Python) is a powerful library used for **scientific computing and numerical operations in Python.** It provides efficient support for large multidimensional arrays and matrices along with mathematical functions. NumPy is widely used in data science, engineering, machine learning, and statistical computing.

NumPy is usually included in the Anaconda distribution. It can also be installed using:

```
pip install numpy
```

Before using NumPy, it must be imported as follows:

```
import numpy as np
```

The alias **np** is commonly used to simplify commands. More details on the numpy usage will be provided in the next chapters. Unlike Python lists, NumPy arrays allow efficient mathematical operations on entire datasets without writing explicit loops.

## Key takeaways

The present chapter introduces both the programming languages making the user comfortable with the environments, the basic structures, naming variable and the basic syntax of naming variables along with the basic arithmetic operations. Here is the summary in concise form:

**MATLAB**

- Matrix-based language, ideal for engineering and numerical computation
- Environment includes Command Window, Workspace, Current Folder, Editor, and Toolboxes
- Syntax: `%` for comments, variables stored as arrays, operators like `+`, `-`, `*`, `/`, with element-wise versions
- Scripts share workspace; functions are modular with isolated scope
- Strength: Specialized toolboxes for domains like signal processing, machine learning, and optimization

**Python**

- General-purpose, open-source, focused on readability and flexibility
- Setup via Python.org or Anaconda; IDEs include IDLE, PyCharm, Spyder, VS Code, Jupyter
- Syntax: `#` for comments, dynamic typing, operators `+`, `-`, `*`, `/`, `**`, `//`, `%`
- Data structures: Lists (mutable), Tuples (immutable), Dictionaries (key-value)
- Functions defined with `def`; supports advanced features like default args and lambdas
- Strength: Rich ecosystem of libraries (NumPy, Pandas, Matplotlib, Scikit-learn)

Having established the fundamental environments, syntax, and operations of MATLAB and Python, the subsequent chapters will progress from this foundational knowledge toward advanced applications. Readers will move from basic programming constructs into deeper explorations of data structures, specialized libraries, and domain-specific problem solving, thereby developing the expertise required to apply both languages effectively in complex scientific and engineering contexts.

## Exercise

### Section 1: MATLAB exercise

1 Display the text **Hello World** using the **display('Hello World')** or **disp('Hello World')** as MATLAB command in the console.

2 Type the following commands in MATLAB and observe the Workspace:

a = 10;

b = 90;

a + b

What is the effect of using a semicolon (;) at the end of a statement?

3 Clear the Command Window using the appropriate MATLAB command **clc**.

4 Display the variables currently available in the workspace using:

– who

– whos

Explain the difference between the two commands.

5 Evaluate the following and observe the difference between round, floor, and ceil:

a) 2 + round(6/7) * 3

b) 3 * floor(7/3)

c) 14 + ceil(15/6)

6 Evaluate the following expressions and state the operator precedence applied:

a) 5 * 6 – 17 * 4

b) 12 * 3^2 / 4

c) 16 / 4 + 20 / 5 – 6

d) sind(90)^2

e) cosd(45)^2 + sind(45)^2

7 Define complex numbers:

a = 3 + 4i;

b = 4 + 6i;

Perform the following operations:

a) a + b

b) a * b

c) abs(a)

d) isreal(a)

8 Evaluate the following expression:

log(20) / log(2)

What does this value represent?

9 Verify whether the following expression evaluates to zero:

$$\frac{(8^2)}{(4+92)} - \frac{8x4}{3(10+6)}$$

10 Create a sequence of numbers from 1 to 10 and store it in a variable x. For example: x=1:10 and compute:
   a) Sum of all elements using sum command
   b) Mean of the elements using mean command
   c) Square of each element using .^ for power
   Interpret the results.

## Section 2: Python exercise

1 Print the text "Hello World" using the print() function. Observe the difference between single quotes and double quotes.
2 Type the following in Python and observe the output:
   a = 10
   b = 90
   a + b
   What is the difference between writing a statement with and without print()?
3 Display all currently defined variables using:
   dir()
   Compare this with using type() for individual variables.
4 Anil ordered a pizza for a small group. He has 11 slices of pizza and 4 friends.
   a) Write a line of code using Standard Division (/) to find out exactly how much pizza each person gets if he cut the slices into pieces.
   b) Write a line of code using Floor Division (//) to find out how many full, whole slices each person gets.
   c) Write a line of code using the Modulo (%) operator to find out how many slices are left over in the box.
5 Evaluate the following using the math module (import math):
   a) 2 + round(6/7) * 3
   b) 3 * math.floor(7/3)
   c) 14 + math.ceil(15/6)
6 Evaluate the following expressions and explain operator precedence:
   a) 5 * 6 – 17 * 4
   b) 12 * 3**2 / 4
   c) 16 / 4 + 20 / 5 – 6
   d) (math.sin(math.radians(90)))**2
   e) (math.cos(math.radians(45)))**2 + (math.sin(math.radians(45)))**2
7 Define complex numbers:
   a = 3 + 4j

b = 4 + 6j
Perform:
a) a + b
b) a * b
c) abs(a)
d) isinstance(a, complex)

8 Evaluate:
import math
math.log(20)/math.log(2)
What does the value represent?

9 Verify whether the following expression evaluates to zero:

$$\frac{(8^2)}{(4+92)} - \frac{8x4}{3(10+6)}$$

10 Create a list from 1 to 10 and compute:
a) Sum of all elements
b) Mean of the elements
c) Square of each element (use list comprehension)
Interpret the results.

### MATLAB solutions

1. Both **display('Hello World')** and **disp('Hello World')** are used to display in message on console.
2. a + b = 100
   Semicolon suppresses output display.
3. clc
4. who → lists variable names.
   whos → lists names, size, bytes, and class information.
5. a) round(6/7) = 1 → 2 + 13 = 5
   b) *floor*(7/3) = 2 → 32 = 6
   c) ceil(15/6) = 3 → 14 + 3 = 17
6. a) 56 – 174 = 30 – 68 = −38
   b) 3^2 = 9 → 12*9/4 = 108/4 = 27
   c) 16/4 + 20/5 – 6 = 4 + 4 – 6 = 2
   d) sind(90) = 1 → 1^2 = 1
   e) cosd(45)^2 + sind(45)^2 = 0.5 + 0.5 = 1.0000
7. a) (3+4i) + (4+6i) = 7 + 10i
   b) (3+4i)(4+6i) = 12 + 18i + 16i + 24i^2 = 12 + 34i – 24 = −12 + 34i

c) abs(a) = 5
d) isreal(a) = 0 (false)

8. log(20)/log(2) ≈ 4.3219
This represents log base 2 of 20.
9. Expression (8^2)/(4 + 92) – (8*4)/(3*(10 + 6))=0
10. x = 1:10
a) sum(x) = 55
b) mean(x) = 5.5000
c) x.^2 = 1 4 9 16 25 36 49 64 81 100
**Interpretation:** The sum and mean describe central tendency; element-wise squaring uses the dot operator (.^) in MATLAB.

## Python solutions

1. Both print('Hello World') and print("Hello World") print the same output **Hello World**
2. a + b = 100
Without print(), output appears only in interactive mode.
3. dir() lists all variables currently defined in the environment.
type() shows the data type of a specific variable.
4. slices = 11
friends = 4
# 1. Exact amount (Standard Division)
# This will give a decimal answer.
print("Exact amount per person:", slices / friends)
# 2. Whole slices (Floor Division)
# This removes the decimal to show only full slices.
print("Whole slices per person:", slices // friends)
# 3. Leftover slices (Modulo)
# This shows the remainder.
print("Slices left in the box:", slices % friends)
**Output:**
Exact amount per person: 2.75
Whole slices per person: 2
Slices left in the box: 3
5. a) 5
b) type the command as:
import math
3 * math.floor(7/3)
Output: 6
c) 17

6. a) −38
   b) 27.0
   c) 2.0
   d) 1.0
   e) 1.0
7. a) 7+10j
   b) −12+34j
   c) 5.0
   d) True
8. ≈ 4.3219 (log base 2 of 20)
9. (8**2)/(4 + 92) − (8*4)/(3*(10 + 6))
10. Define and print x using the following:
    x=list(range(1,11))
    print(x)
    a) sum(x) = 55
    b) sum(x)/len(x) = 5.5
    c) [i**2 for i in x] = [1, 4, 9, 16, 25, 36, 49, 64, 81, 100]
       Interpretation: Python uses list structures and list comprehension for element-wise operations.

    Another way to do this is to use **numpy**

```
import numpy as np
# Create a vector (as a list, then convert to numpy array)
x=list(range(1,11))
v = np.array(x)
print(np.sum(v))
print(np.mean(v))
S= v ** 2
print(S)
```

Following is the output:

```
55
5.5
[ 1 4 9 16 25 36 49 64 81 100]
```

## Computational practice questions with short answers

Q1 What is the output of the following MATLAB command?

```
a = 4; b = 3;
result = a^2 + b^2
```

Q2 What will the following Python code output?

```
a = 7
b = 2
print(a // b, a % b)
```

Q3 Evaluate the following MATLAB expression and state the result:

```
x = (5 + 3) * 2 – 4 / 2
```

Q4 What is the output of the following Python code?

```
x = 2 ** 10
print(x)
```

Q5 Find the result of the following MATLAB command:

```
a = mod(17, 5)
```

Q6 What will the following Python code output?

```
import math
print(math.sqrt(225))
```

Q7 What is the result stored in z after executing the following MATLAB code?

```
x = 10;
y = 3;
z = x * y – x / y
```

Q8 What will be the output of the following Python code?

```
x = 5
x = x + 3
x *= 2
print(x)
```

Q9 What does the following MATLAB code display?

```
a = 100;
b = sqrt(a);
c = b^3;
disp(c)
```

Q10 Predict the output of the following Python code:

```
a = 3
b = 4
c = (a**2 + b**2) ** 0.5
print(c)
```

Q11 What is the output of the following MATLAB commands?

```
x = 8;
y = x / 0
```

Q12 What will the following Python code output?

```
x = 10
y = 3
print(x / y)
print(round(x / y, 2))
```

Q13 What is the output of the following MATLAB code?

```
x = log(exp(5))
```

Q14 What will the following Python code output?

```
import math
print(math.log(math.exp(7)))
```

Q15 Evaluate the following MATLAB expression:

```
a = floor(7.9)
b = ceil(7.1)
c = round(7.5)
```

Q16 What will the following Python code output?

```
import math
print(math.floor(9.8))
print(math.ceil(9.2))
print(round(9.5))
```

Q17 What is the output of the following MATLAB command?

```
x = abs(-45) + abs(15)
```

Q18 What will the following Python code output?

```
import math
x = math.log10(1000) + math.log10(100)
print(x)
```

Q19 What is the result of the following MATLAB code?

```
v = [1 2 3 4 5];
s = sum(v);
m = mean(v);
disp([s, m])
```

Q20 What will the following Python code output?

```
v = [10, 20, 30, 40, 50]
print(sum(v))
print(sum(v) / len(v))
```

Q21 What is the output of the following MATLAB code?

```
A = [1 2; 3 4];
B = A * 2;
disp(B)
```

Q22 What will the following Python code output?

```
A = [[1, 2], [3, 4]]
B = [[A[i][j] * 2 for j in range(2)] for i in range(2)]
print(B)
```

Q23 What is the final value of result in the following MATLAB code?

```
x = 9;
y = 4;
result = sqrt(x) * y^2 - mod(x, y)
```

Q24 What will the following Python code compute and print?

```
import math
r = 7
area = math.pi * r**2
circumference = 2 * math.pi * r
print(round(area, 2))
print(round(circumference, 2))
```

Q25 What is the output of the following MATLAB code?

```
x = 2;
for i = 1:4
x = x * 2;
end
disp(x)
```

**Solutions:**

| | | | | |
|---|---|---|---|---|
| Q1) 25 | Q2) 3 1 | Q3) 14 | Q4) 1024 | Q5) 2 |
| Q6) 15.0 | Q7) 26.6667 | Q8) 16 | Q9) 1000 | Q10) 5.0 |
| Q11) Inf | Q12) 3.3333333333333335, 3.33 | Q13) 5 | Q14) 7.0 | Q15) 7, 8, 8 |
| Q16) 9, 10, 10 | Q17) 60 | Q18) 5.0 | Q19) 15, 3 | Q20) 150, 30.0 |
| Q21) [2 4; 6 8] | Q22) [[2, 4], [6, 8]] | Q23) 47 | Q24) 153.94, 43.98 | Q25) 32 |

CHAPTER 2

# MATLAB fundamentals: vectors and arrays in scientific calculations

After discussing the environment setup and basic MATLAB commands in the first chapter, this chapter advances the reader's understanding by introducing the fundamental concepts of vectors and arrays. Vectors and arrays form the core building blocks of MATLAB and are essential for performing a wide range of mathematical operations.

This chapter emphasizes the importan.ce of defining, manipulating, and working efficiently with vectors and arrays. Readers will learn how to create, modify, and manage arrays using various built-in MATLAB functions and commands, thereby developing a strong foundation for computational and mathematical applications.

## Defining vectors

Vectors are one-dimensional arrays used to represent quantities such as time series, measurements, or coefficients. A vector can be defined in either the row or column form. A vector can be defined as a collection of items combined together stored under the name of a variable. To make the different elements clubbed together, the items should be added in a square bracket either separated by a space or a comma. This creates a **row vector.** Instead, if the elements are clubbed together separated by a semi colon creates a column vector.

Here is an example collecting the data values with the variable name r and c in the row and column form, respectively:

- **Row vector:**

```
r = [11 62 35 64 50]
```

or

```
r = [11, 62, 35, 64, 50]
```

 | HTTPS://DOI.ORG/10.1515/9783112207758-002

**Output:** r =
11 62 35 64 50
– **Column vector:**

```
c = [41; 72; 93; 42; 15]
```

**Output:** c =
41
72
93
42
15
Another way to define a vector is to define with the colon operator when the elements are distributed with a common difference d. The syntax to define a vector using colon is:

```
Initial: common difference: terminal value
```

The value of the increment (difference) can be either positive or negative, depending on the requirement. By default, the increment is +1, and it does not need to be explicitly specified. The examples in Table 2.1 illustrate different ways to define vectors in MATLAB, along with their corresponding outputs.

TABLE 2.1 Vector creation using the colon operator.

| Requirement | Command | Output |
|---|---|---|
| Define a vector with elements from 2 to 13 | x=2:13 | x = 2 3 4 5 6 7 8 9 10 11 12 13 |
| Define a vector with **even** elements from 1 to 20 | a=2:2:20 | a = 2 4 6 8 10 12 14 16 18 20 |
| Define a vector with **odd** elements from 1 to 20 | b=1:2:19 | b = 1 3 5 7 9 11 13 15 17 19 |
| Define a vector with the **multiples of 3** elements from 1 to 20 | c=3:3:18 | c = 3 6 9 12 15 18 |
| Define a vector in **reverse order** elements from 20 to 1 | d=20:−1:1 | d = 20 19 18 17 16 15 14 13 12 11 10 9 8 7 6 5 4 3 2 1 |
| Define a vector in **reverse order** elements from 20 to 1 with a **common difference of 3** | e=20:−3:1 | e = 20 17 14 11 8 5 2 |

Another way to define a vector with the initial and terminal values along with the point it need to be partitioned is **linspace.**

**The linspace(a, b, n) function generates n equally spaced points between a and b.**

**Note:** if n is omitted in linspace(a, b), MATLAB defaults to 100 **points.**

**Example 1:**

```
linspace(2, 13, 6)
```

**Output:**
2.0000 4.2000 6.4000 8.6000 10.8000 13.0000

**Example 2:**

```
linspace(0, 1, 11)
```

**Output:**
0 0.1000 0.2000 0.3000 0.4000 0.5000 0.6000 0.7000 0.8000 0.9000 1.0000
Once a vector is generated, there is a need to work on its specifications and manipulating the elements. Let us now focus on the commands required to perform the computations on a vector.

### Accessing vector elements

Element of the vectors can be accessed by calling from its position. For a vector x, the elements in it can be called as x(n) where n refers to the position of the element. The first position is 1. Consider an example of the first 10 odd numbers.

**Command:**

```
X=1:2:20
```

**Output:**

X =

1 3 5 7 9 11 13 15 17 19

Here are the commands along with the extraction required in Table 2.2.

TABLE 2.2 Examples of accessing and extracting elements from a vector.

| Requirement | Command | Output |
|---|---|---|
| Extract the element at the third position of the vector | X(3) | 5 |
| Extract the last element of the vector | X(end) | 19 |
| Extract the first four elements of the vector | X(1:4) | 1 3 5 7 |
| Extract the last four elements of the vector | X(end-3:end) | 13 15 17 19 |
| Extract the elements at second, fourth, and seventh position of the vector | X([2,4,7]) | 3 7 13 |
| Extract the elements at the even positions of the vector | X(2:2:end) | 3 7 11 15 19 |

## Mathematical operations within a vector

A vector can be operated with the mathematical operations and results in the required output. There are many commands that work on the elements of the vectors as discussed in Table 2.3.

TABLE 2.3 Basic mathematical and structural operations on a vector.

| Requirement | Command | Output |
|---|---|---|
| Create a vector V with elements as 11, 23, 54, 1, 10 | V=[11, 23, 54, 1, 10] | V =<br>11 23 54 1 10 |
| Find the length (number of elements) of the vector V | length(V) | ans =<br>5 |
| Find the size of the vector | size(V) | ans =<br>1 5 |
| Find the sum of the elements of the vector V | sum(V) | ans =<br>99 |
| Find the average of the elements of the vector V | mean(V) | ans =<br>19.8000 |

TABLE 2.3 Basic mathematical and structural operations on a vector (*cont.*)

| Requirement | Command | Output |
|---|---|---|
| Find the transpose of the elements of the vector V | transpose(V) | ans =<br>11<br>23<br>54<br>1<br>10 |
| Find the product of the elements of the vector V | prod(V) | ans =<br>136620 |
| Assign a value 0 to the fifth element of a vector V | V(5)=0 | V =<br>11 23 54 1 0 |
| Add 5 to the fourth element of the vector V | 5+V(4) operation does not modify the vector V because the result is not reassigned.<br>To get it visible in the vector the command is:<br>V(4)=5+V(4) | V =<br>11 23 54 6 0 |
| Subtract 3 from the last element of the vector V | V(end)=V(end)-3 | V =<br>11 23 54 6 −3 |
| Add three new elements to the matrix as 10, 20, 40 | V=[V 10 20 40] | V =<br>11 23 54 6<br>−3 10 20 40 |
| Add an element 100 at the 12th position and verify that the unavailable elements are assigned zeros in the vector V | V(12)=100 | V =<br>11 23 54 6<br>−3 10 20 40<br>0 0 0 100 |
| Flip the elements of the vector from left to right | V=fliplr(V) | V=<br>100 0 0 0 40<br>20 10 −3 6<br>54 23 11 |

### Mathematical operations with vectors

Once learned about the mathematical operations in a vector for its element. Let's learn about the operations that includes two or more vectors.

The basic operations between the two vectors include their sum, difference, and product. Consider the following two example vectors to learn the manipulation.

Let the two vectors are **S=1:3:12** and **R=2:2:8.**

**This results in the values as:**

S = 1 4 7 10
R = 2 4 6 8

Table 2.4 explains the operations on the two vectors that can be extended in a similar way for more than two vectors.

TABLE 2.4 Arithmetic operations between two vectors.

| Requirement | Command | Output |
|---|---|---|
| Find the sum of vectors S and R in two different ways | S+R | ans =<br>3 8 13 18 |
| Find the difference of vectors S and R | S-R | ans =<br>−1 0 1 2 |
| Find the product of vectors S and R | S*R | Error using *<br>Inner matrix dimensions must agree. |
| **Explanation:** | The vectors S and R are both row vectors. Standard matrix multiplication requires the inner dimensions to match. Since both vectors are 1×4, the multiplication S*R is not defined. In order to multiply the elements of S and R, instead of S*R the correct formula is S.*R, this is called element by element multiplication.<br>**Output with command:**<br>**S.*R**<br>ans =<br>2 16 42 80 | |

TABLE 2.4 Arithmetic operations between two vectors (*cont.*)

| Requirement | Command | Output |
|---|---|---|
| Find the value of $2^R$ | 2^R | Error using ^<br>Inputs must be a scalar and a square matrix.<br>To compute elementwise POWER, use POWER (.^) instead. |
| **Explanation:** | A scalar cannot be raised to a vector using the matrix power operator (^). To compute powers element-by-element, the dot operator (.^) must be used. Using 2.^R results in 2 raise to the values with elements of R. with the **output** as:<br>ans =<br>4 16 64 256 | |

**Note:** For standard matrix multiplication, the inner dimensions must agree. A matrix with size **(n,p)** can be multiplied by another matrix of size **(p,r)** so that the resultant matrix is **(n,r).**

### Concatenation of vectors

Concatenation is the process of joining small variables together to create a larger one. This is essential when building complex datasets or adding new results to existing arrays:

- **Horizontal concatenation:** To join two vectors side-by-side, use a space or comma within square brackets.
  **Command:** C = [A, B] or C = [A B]
- **Vertical concatenation:** To stack one vector on top of another, use a semicolon. Note that the number of columns in both vectors must be identical:
  **Command:** C = [A; B]
- **Example:**
  If A = [1, 2] and B = [3, 4], then [A, B] results in [1, 2, 3, 4].

## Working with arrays

Arrays are extensions of vectors into two or more dimensions. Arrays are essential in scientific calculations such as solving systems of equations or representing experimental data. It represents the elements or items arranged in both the row as well as the column form. Arrays can be created defining the row elements separated by the space or comma followed by the semicolon to define the elements of the next row.

User should define the rows with the similar number of elements as these created the column elements of the array. The syntax can be defined as follows:

```
Array_name=[first row ; second row; third row, ..., last row]
```

Let's learn with help of an example to create a matrix defined as:

$$A = \begin{bmatrix} 1 & 3 & 6 \\ 11 & 0 & 9 \\ 3 & 5 & 12 \end{bmatrix}$$

The command to create this matrix A is:

```
A = [1 3 6; 11 0 9; 3 5 12];
```

The matrix can be defined using the colon operator when the elements are continuous or can be defined as a predefined vector. Here is another example to demonstrate the same:

$$B = \begin{bmatrix} 1 & 2 & 3 \\ 4 & 6 & 8 \\ 3 & 5 & 12 \end{bmatrix}$$

```
B=[1:3; 4:2:8; [3,5,12]]
```

Here is another example using the linspace, the trigonometric function and a predefined vector as the third row in the array:

```
A=[1, 23, 5, 67]
C=[linspace(1,2,4); sin([pi,pi/2,pi/4,pi]);A]
```

**Output:**

```
C =

1.0000    1.3333    1.6667    2.0000
0.0000    1.0000    0.7071    0.0000
1.0000   23.0000    5.0000   67.0000
```

### Built-in matrix generators

MATLAB provides several commands that assist user to create some commonly known matrices. The commands along with the examples to create matrix with all elements as zeros, all elements as ones and identity matrix are given in Table 2.5.

TABLE 2.5 Creation of standard matrices using built-in MATLAB functions.

| Requirement | Command | Output |
|---|---|---|
| Create a matrix X with 3 rows and 4 columns using the function **zeros(r,c)** where r denoted rows and c denote column | X=zeros(3,4) | X =<br>0 0 0 0<br>0 0 0 0<br>0 0 0 0 |
| Create a matrix Y with 5 rows and 6 columns using the function **ones(r,c)** where r denoted rows and c denote column | Y=ones(5,6) | Y =<br>1 1 1 1 1 1<br>1 1 1 1 1 1<br>1 1 1 1 1 1<br>1 1 1 1 1 1<br>1 1 1 1 1 1 |
| Create a square matrix Z of size 3 that must be an identity matrix using the function **eye(n)** where n is common for both rows and column | Z=eye(3) | Z =<br>1 0 0<br>0 1 0<br>0 0 1 |

### Accessing matrix elements

Once a matrix is defined, the elements to the matrix can be accessed using the row number and the column number in the variable that stores the matrix. Thus, the element can be extracted using the appropriate row and column numbers in the command:

```
A(row, column)
```

Let's consider an example matrix A created earlier as

$$A = \begin{bmatrix} 1 & 3 & 6 \\ 11 & 0 & 9 \\ 3 & 5 & 12 \end{bmatrix}$$

The command for its creation is **A** = [1 3 6; 11 0 9; 3 5 12]

The element on the second row and the first column can be accessed using the command:

```
A(2,1)
```

**Output:** 11

Similarly, to get the element at the third row and the second column, the command is:

```
A(3,2)
```

**Output:** 5

To get the elements for the whole row or a column, the colon is used. For example, to extract the second row of the matrix A, the command is:

```
A(2,:)
```

**Output:** 11 0 9

Some other relevant examples are listed in Table 2.6 shown on a matrix F:
Define matrix:

$$F = \begin{bmatrix} 11 & 30 & 38 \\ 0 & 14 & 46 \\ 52 & 10 & 17 \\ 10 & 5 & 11 \end{bmatrix}$$

Using command:

```
F = [11 30 38; 0 14 46;52 10 17;10 5 11]
```

TABLE 2.6 Examples of accessing and manipulating matrix elements.

| Requirement | Command | Output |
|---|---|---|
| Elements of the third row with the first and third column | F(3,[1,3]) | ans =<br>52 17 |
| Element with command F(6) | F(6) | ans =<br>14 |
| All elements of the matrix as a column | F(:) | ans =<br>11<br>0<br>52<br>10<br>30<br>14<br>10<br>5<br>38<br>46<br>17<br>11 |
| Transpose the row element to column of matrix F and save it as G | G=F' | G =<br>11 0 52 10<br>30 14 10 5<br>38 46 17 11 |
| Use relevant command to extract a submatrix from matrix G as<br>14 5<br>46 11 | G(2:3,[2,4]) | ans =<br>14 5<br>46 11 |

TABLE 2.6 Examples of accessing and manipulating matrix elements (*cont.*)

| Requirement | Command | Output |
|---|---|---|
| All elements of column two from matrix G | G(:,2) | ans =<br>0<br>14<br>46 |
| Delete the first row of the matrix F using the empty brackets | F(1,:)=[] | F =<br>0 14 46<br>52 10 17<br>10 5 11 |
| Delete the second column of the updated matrix F | F(:,2)=[] | F =<br>0 46<br>52 17<br>10 11 |
| Delete the first and third row of the updated matrix F | F([1,3],:)=[] | F =<br>52 17 |

The indexing operations can be summarized in Table 2.7.

TABLE 2.7 Summary of common matrix indexing syntax.

| Syntax | Description | Result type |
|---|---|---|
| F(:, n) | All rows of column n | Column vector |
| F(n, :) | All columns of row n | Row vector |
| F(m:n, p:q) | Intersection of row range m:n and column range p:q | Sub-matrix |

### Distinguishing size, length, and numel

Understanding the dimensions of the data is critical for scientific calculations. Table 2.8 compares the three primary commands used to check the array specifications. The first two commands as size and length are already discussed for the vectors. To present the understanding, here is an array consider, for example, as:

$$B = \begin{bmatrix} 1 & 3 & 9 \\ 2 & 4 & 6 \\ 5 & 1 & 7 \\ 10 & 8 & 15 \end{bmatrix}$$

Created using command B=[1 3 9; 2 4 6; 5 1 7; 10 8 15].

TABLE 2.8 Comparison of commands for determining array dimensions.

| Requirement | Command | Output |
|---|---|---|
| Find dimensions | size(B) | ans =<br>4 3 |
| Find longest dimension | length(B) | ans =<br>4 |
| Find total elements | numel(B) | ans =<br>12 |

## Array operations

There are many commands associated with the matrix manipulation. The basic operations include the sum, average, mean, median, determinant, trace, rank, diagonal, eigenvalue, inv. Other than these there are some special operations required in the scientific computation that include specific element-wise multiplication. Table 2.9 demonstrates the basic matrix operations.

Let us consider a matrix to demonstrate the matrix operations, using command: W=[2 3 0 8;34 12 11 4; 41 27 13 7;17 13 19 9]

$$W = \begin{bmatrix} 2 & 3 & 0 & 8 \\ 34 & 12 & 11 & 4 \\ 41 & 27 & 13 & 7 \\ 17 & 13 & 19 & 9 \end{bmatrix}$$

Table 2.9 provides the basic mathematical operations that can be performed on a matrix. Here are some more commands to do some specific operations. Out of which the common operation includes the element-by-element operations and solving the set of equations

TABLE 2.9 Common mathematical operations on matrices.

| Requirement | Command | Output |
|---|---|---|
| Sum of all the elements of matrix W | sum(W(:)) | ans =<br>220 |
| Sum of the elements of matrix W row wise | sum(W,1) | ans =<br>94 55 43 28 |
| | sum(W) | ans =<br>94 55 43 28 |
| Sum of the elements of matrix W column wise | sum(W,2) | ans =<br>13<br>61<br>88<br>58 |
| Average of all the elements of matrix W | mean(W(:)) | ans = 13.7500 |
| Average of the elements of matrix W row wise | mean(W) | ans =<br>23.5000 13.7500 10.7500 7.0000 |
| | mean(W,1) | ans =<br>23.5000 13.7500 10.7500 7.0000 |
| Average of the elements of matrix W column wise | mean(W,2) | ans =<br>3.2500<br>15.2500<br>22.0000<br>14.5000 |
| Median of all the elements of matrix W | median(W(:)) | ans =<br>11.5000 |
| Median of the elements of matrix W row wise | median(W) | ans =<br>25.5000 12.5000 12.0000 7.5000 |
| Median of the elements of matrix W column wise | median(W,2) | ans =<br>2.5000<br>11.5000<br>20.0000<br>15.0000 |
| Find determinant of the matrix W | det(W) | ans =<br>−44302 |
| Find trace of the matrix W | trace(W) | ans =<br>36 |
| Find rank of the matrix W | rank(W) | ans =<br>4 |

TABLE 2.9 Common mathematical operations on matrices (*cont.*)

| Requirement | Command | Output |
|---|---|---|
| Find super-diagonal value of the matrix W | diag(W,1) | ans =<br>3<br>11<br>7 |
| Find sub-diagonal value of the matrix W | diag(W,−1) | ans =<br>34<br>27<br>19 |
| Find eigenvalues of the matrix W | eig(W) | ans =<br>44.9789 + 0.0000i<br>−0.5319 +11.1426i<br>−0.5319 −11.1426i<br>−7.9150 + 0.0000i |
| Find the eigenvalue as well as the eigen vectors of W | [a,b]=eig(W) | a =<br>0.1384 + 0.0000i −0.0142 – 0.3702i<br>−0.0142 + 0.3702i −0.3850 + 0.0000i<br>0.4376 + 0.0000i −0.3162 + 0.2598i<br>−0.3162 – 0.2598i 0.8286 + 0.0000i<br>0.6737 + 0.0000i −0.0914 + 0.5281i<br>−0.0914 – 0.5281i −0.3707 + 0.0000i<br>0.5792 + 0.0000i 0.6387 + 0.0000i<br>0.6387 + 0.0000i 0.1665 + 0.0000i<br>b = 44.9789 + 0.0000i 0.0000 +<br>0.0000i 0.0000 + 0.0000i 0.0000 +<br>0.0000i<br>0.0000 + 0.0000i −0.5319 +11.1426i<br>0.0000 + 0.0000i 0.0000 + 0.0000i<br>0.0000 + 0.0000i 0.0000 + 0.0000i<br>−0.5319 −11.1426i 0.0000 + 0.0000i<br>0.0000 + 0.0000i 0.0000 + 0.0000i<br>0.0000 + 0.0000i −7.9150 + 0.0000i |

### Solving systems of linear equations

One of the most powerful applications of arrays in MATLAB is solving systems of linear equations. A system of equations can be represented in matrix form as AX = B, where:

**A** is the coefficient matrix.

**X** is the vector of unknowns (x, y, z).

**B** is the constant vector.

Consider the following system:

$$x + 2y + 5z = 9$$

$$2x - 3y + 2z = 9$$

$$9y + 8z = 7$$

To solve this in MATLAB, it requires to define the coefficient matrix A and the constant vector B. Note that for the third equation, the coefficient of y is 0.

Following are the commands required to define the matrix and the column vector as required to solve this example of the system of equation:

```
A = [1 2 5; -4 1 2; 0 9 8];
B = [12; 13; 1];                % using semicolons to create a column vector.
```

**Solution command:**

```
X = A \ B                %This uses the backslash operator for matrix division.
```

Another way to solve this system is using the inv command to find the inverse as:

```
X=inv(A)*B
```

Here is the output for the commands along with the solution:

```
X =
-2.0317
-3.6984
4.2857
```

Be noted that MATLAB recommends using the backslash operator (A\B) because it is numerically more stable and computationally efficient.

### Exercise

1. **Variable substitution:** Define variables a = 2.5 and b = 4.2. Perform the following operations on the variables: a*b, a^b, b/a, and a+b.
2. **Colon operator:** Create a variable **k** as a row vector where the first element is 5, the last element is 45, and the elements increase by an increment of 5.
3. **Linear spacing:** Create a row vector **h** with **six equally spaced elements** where the first element is 10 and the last element is 50.
4. **Prime numbers:** Define a vector V containing the prime numbers less than or equal to 10 using primes(10).
5. **Data filtering:** Create a vector t from 0 to 10 with 50 points using linspace. Create a second vector y = sin(t). Write a command to extract all values of y that are positive.
6. **Logical operations:** Given the vector V = [11, 23, 54, 1, 10], write a single line of code using logical indexing to double the value of only the elements that are greater than 20.
7. **Sorting and indexing:** Create a vector X = [1, 23, 29, 34, 18, 10]. Sort the elements of X in ascending order using the sort command, then use the command X(X > 10) to find all elements greater than 10.

#### Section B: array manipulation and concatenation

8. **Horizontal and vertical joining:** Define any two vectors of size 10, named Y and Z. Perform the following:
   a) Join them side-by-side using the command Q = [Y, Z] (horizontal concatenation).
   b) Stack them using the command Q = [Y; Z] (vertical concatenation).
9. **Matrix modification:** Use the command R = rand(4) to define a 4 times 4 random matrix. Write a command to replace the entire second column of R with zeros.
10. **Special matrix generators:** Using the zeros, ones, and eye commands, create the following arrays with a single command:
    a) A 3 times 6 matrix, where the left half is an identity matrix and the right half is also an identity matrix.
    b) A 2 times 5 matrix, where the first three columns are zeros and the last two columns are ones.
11. **Sub-matrix extraction:** Using the matrix F defined in the chapter:
    a) Create a row vector Ua containing the first three elements of the first row followed by the last three elements of the third row.
    b) Create a column vector Ub that concatenate the first, second, and third columns of F

## Section C: advanced computational structures

1. **Systems of equations:** Represent the following system in matrix form (AX=B) and solve for X using the backslash operator:

   $x + 2y + 3z = 12$

   $-4x + y + 2z = 13$

   $9y + 8z = 1$

2. **Tridiagonal matrix:** Create the following tridiagonal matrix without typing elements explicitly:

$$\begin{bmatrix} 4 & 2 & 0 & . & 0 \\ 1 & 4 & 1 & 0 & 0 \\ 0 & \ddots & \ddots & \ddots & 0 \\ 0 & 0 & 1 & 4 & 1 \\ 0 & 0 & 0 & 2 & 4 \end{bmatrix}$$

3. **Pentadiagonal matrix:** Construct the pentadiagonal matrix defined below. Utilize the **diag** command for each offset diagonal, as discussed in the chapter

$$\begin{bmatrix} 18 & 6 & 0 & 0 & 0 & 0 & 0 & 0 \\ 10 & 3 & 1 & 0 & 0 & 0 & 0 & 0 \\ 1 & 11 & 11 & 1 & 0 & 0 & 0 & 0 \\ 0 & 1 & 11 & 11 & 1 & 0 & 0 & 0 \\ 0 & 0 & \ddots & \ddots & \ddots & \ddots & 0 & 0 \\ 0 & 0 & 0 & 1 & 11 & 11 & 1 & 0 \\ 0 & 0 & 0 & 0 & 0 & 1 & 3 & 10 \\ 0 & 0 & 0 & 0 & 0 & 0 & 6 & 8 \end{bmatrix}$$

## Computational practice questions with short answers

Q1. What is the output of the following MATLAB code?

```
v = [2 4 6 8 10];
disp(v(3))
```

Q2. What will the following MATLAB code produce?

```
v = 1:2:10;
disp(v)
```

Q3. Find the output of the following MATLAB code:
```
v = linspace(0, 1, 5);
disp(v)
```
Q4. What is the result of the following MATLAB code?
```
a = [3 6 9 12 15];
b = a / 3;
disp(b)
```
Q5. What is the output of the following MATLAB code?
```
v = [10 20 30 40 50];
disp(length(v))
disp(sum(v))
disp(mean(v))
```
Q6. What does the following MATLAB code output?
```
v = [5 3 8 1 9 2];
disp(max(v))
disp(min(v))
disp(sort(v))
```
Q7. What is the output of the following MATLAB code?
```
a = [1 2 3 4];
b = [5 6 7 8];
c = a + b;
d = a .* b;
disp(c)
disp(d)
```
Q8. What will the following MATLAB code output?
```
a = [10 20 30 40];
b = [2 4 5 8];
c = a ./ b;
disp(c)
```
Q9. What is the result of the following MATLAB code?
```
v = [1 2 3 4 5];
result = v .^ 2;
disp(result)
```
Q10. Predict the output of the following MATLAB code:
```
x = [0 1 2 3];
y = 2.^x;
disp(y)
```
Q11. What does the following MATLAB code compute?
```
a = [4 9 16 25 36];
b = sqrt(a);
disp(b)
```

Q12. What is the output of the following MATLAB code?

```
v = [1 2 3 4 5 6];
disp(v(2:5))
disp(v(end))
disp(v(1:2:end))
```

Q13. What will the following MATLAB code output?

```
A = [1 2 3; 4 5 6; 7 8 9];
disp(size(A))
disp(A(2,3))
```

Q14. What is the result of the following MATLAB code?

```
A = [1 2; 3 4];
B = [5 6; 7 8];
C = A + B;
disp(C)
```

Q15. What does the following MATLAB code produce?

```
A = [1 2; 3 4];
B = A * A;
disp(B)
```

Q16. What is the output of the following MATLAB code?

```
A = [1 2; 3 4];
disp(det(A))
disp(trace(A))
```

Q17. What will the following MATLAB code produce?

```
A = zeros(3,3);
A(2,2) = 5;
disp(A)
```

Q18. What is the output of the following MATLAB code?

```
A = [4 7; 2 6];
B = inv(A);
disp(B)
```

Q19. What will the following MATLAB code compute and display?

```
t = 0:0.5:2;
v = 9.8 * t;
disp(v)
```

Q20. What does the following MATLAB code compute?

```
r = [1 2 3 4 5];
Area = pi * r.^2;
disp(round(Area, 2))
```

### Solutions

| | | | | |
|---|---|---|---|---|
| Q1) 6 | Q2) 1 3 5 7 9 | Q3) 0 0.25 0.5 0.75 1 | Q4) 1 2 3 4 5 | Q5) 5, 150, 30 |
| Q6) 9, 1, 1 2 3 5 8 9 | Q7) 6 8 10 12, 5 12 21 32 | Q8) 5 5 6 5 | Q9) 1 4 9 16 25 | Q10) 1 2 4 8 |
| Q11) 2 3 4 5 6 | Q12) 2 3 4 5, 6, 1 3 5 | Q13) 3 3, 6 | Q14) 6 8; 10 12 | Q15) 7 10; 15 22 |
| Q16) −2, 5 | Q17) [0 0 0; 0 5 0; 0 0 0] | Q18) 0.6 −0.7; −0.2 0.4 | Q19) 0 4.9 9.8 14.7 19.6 | Q20) 3.14 12.57 28.27 50.27 78.54 |

## Mini project

Scientific Data Analysis of Projectile Motion Using Vectors and Arrays in MATLAB

### Project overview

In this mini project, you will use MATLAB Vectors and Arrays to simulate, compute, and analyze the motion of a projectile launched at different angles and velocities. You will apply all fundamental concepts of vectors, arrays, element-wise operations, and built-in mathematical functions learned in Chapter 2.

### Project objectives

- Create and manipulate vectors and arrays
- Perform element-wise arithmetic operations
- Use built-in MATLAB functions for scientific computation
- Apply linspace, zeros, ones, size, sum, max, min
- Store results in matrices for organized data analysis
- Compute real-world scientific quantities using arrays

### Scientific background

A projectile launched with initial velocity $v_0$ at angle $\vartheta$ follows these equations:

Horizontal range: $R = \frac{v_0^2 \sin(2\vartheta)}{g}$

Maximum height: $H = \frac{v_0^2 \sin^2(\vartheta)}{2g}$

Time of flight: $T = \frac{2v_0 \sin(\vartheta)}{g}$

where $v_0$ is the initial velocity (m/s), $\vartheta$ is the launch angle (degrees), and $g = 9.8\,\text{m/s}^2$ (acceleration due to gravity).

## Project tasks

### Task 1: Create the angle vector

**Problem:** Create a vector of launch angles from 10° to 80° with a step of 10° and convert them to radians.

**Hint 1:**
- *Use start:step:end syntax to create the angle vector*
- *To convert degrees to radians, multiply by π/180*
- *Use pi keyword in MATLAB for the value of π*

### Task 2: Compute the time of flight

**Problem:** Given initial velocity $v_0 = 50\,\text{m/s}$ and $g = 9.8\,\text{m/s}^2$, compute the Time of Flight for all angles using the formula: $T = \frac{2v_0 \sin(\vartheta)}{g}$.

**Hint 2:**
- *Use element-wise operations since angles_rad is a vector*
- *sin() in MATLAB works directly on vectors – no loop needed*
- *The result T will be a vector of the same size as angles_rad*

### Task 3: Compute the maximum height

**Problem:** Compute the maximum height reached by the projectile at each angle using:

$$H = \frac{v_0^2 \sin^2(\vartheta)}{2g}$$

**Hint 3:**
- *Use .^2 (element-wise power) on sin(angles_rad) – NOT ^2*
- *Using ^2 without the dot will cause a matrix operation error*
- $\sin^2(\vartheta)$ means sin(angles_rad).^2 in MATLAB

### Task 4: Compute the horizontal range

**Problem:** Compute the horizontal range of the projectile at each angle using:

$$R = \frac{v_0^2 \sin(2\vartheta)}{g}$$

**Hint 4:**
- *Notice that Range at 10° = Range at 80° and 40° = 50° – this is a classic symmetry property of projectile motion.*
- *sin(2 * angles_rad) – the 2* multiplies the angle, not the result.*

### Task 5: Build a complete results matrix

**Problem:** Combine all computed vectors – angles, time of flight, max height, range – into a single results matrix and display it neatly.

**Hint 5:**
- *Use semicolons inside [ ] to stack row vectors as matrix rows*
- *size(Results) should return 4 × 8*
- *Each column represents one launch angle*

### Task 6: Statistical analysis of results

**Problem:** Find the maximum range, the angle that gives maximum range, and compute the average height across all angles.

**Hint 6:**
- *Use [value, index] = max(vector) to get both the maximum value and its position*
- *Use that index to find the corresponding angle: angles_deg(idx)*
- *mean() directly computes average of a vector*

### Task 7: Multi-velocity analysis using a matrix

**Problem:** Compute the horizontal range for three different velocities (30, 50, 70 m/s) across all eight angles. Store as a 3 × 8 matrix.

**Hint 7:**
- *Use zeros(3,8) to pre-allocate the matrix before the loop*
- *R_matrix(i,:) means all columns of row i*
- *Higher velocity always gives larger range at the same angle*

CHAPTER 3

# Python fundamentals: lists and arrays in scientific computations

After understanding vectors and arrays in MATLAB in the previous chapter, this chapter introduces similar fundamental concepts in Python. In scientific and numerical computing, Python primarily works with **lists** and **arrays** (using libraries such as NumPy). These structures form the foundation for performing mathematical operations, data analysis, and computational modeling in Python.

This chapter focuses on defining, manipulating, and performing operations on lists and arrays. Readers will learn how to create sequences of numbers, apply mathematical operations, and utilize built-in and library functions to handle numerical data efficiently. By the end of this chapter, the reader will be equipped to perform structured numerical computations in Python.

Before starting with the discussion of the commands dealing with the list and arrays in Python, an important concept of when or when not to use the print command in Python is much required. As in MATLAB each command which is written without a semicolon will give the output it is not the case with Python. Here is a discussion and detail on the expressions and assignments.

## Understanding output behavior in Python

Before beginning the discussion on commands related to lists and arrays in Python, it is essential to clarify an important concept: **when and why the print() command is required.**

In MATLAB, any command written without a semicolon automatically displays its output. However, Python does not follow this rule. The behavior of output display depends on whether the code is written as an expression or a statement, and also on the environment in which Python is being executed.

To understand this clearly, let's differentiate between expressions, statements, and return values.

The following discussion is particularly relevant for users working in Jupyter Notebook, but the fundamental ideas apply to all IDEs.

### Expression

An **expression** is any piece of code that Python evaluates and produces a value.

 | HTTPS://DOI.ORG/10.1515/9783112207758-003

**Examples**

5 + 3

12 * 2

list(range(1, 6))

Each of these produces a value when entered in the cell and run ***Shift*** + ***Enter***. Table 3.1 presents the inputs along with the outputs shown in dark lines for the command in the line above it.

TABLE 3.1 Presentation of the cells of input and output expressions in Jupyter.

```
5+3
8
12*2
24
list(range(1, 6))
[1, 2, 3, 4, 5]
```

### Statement

A **statement** is an instruction that Python executes. It performs an action but does not necessarily produce a displayable value (see Table 3.2).

**Example**

TABLE 3.2 Demonstration of the input and output in Jupyter.

```
x = 10
for i in range(5):
    print(i)
s
0
1
2
3
4
x = list(range(1, 6))
```

The last statement will give no output. See Table 3.3 for the example of printing the variable x with the list of numbers 1–6.

TABLE 3.3 Example of printing the variable with the list of numbers.

```
x = list(range(1, 6))
print(x)
[1, 2, 3, 4, 5]
```

Consider another example to compare the different ways of inputs and the output in Table 3.4.

TABLE 3.4 Different ways to give the input and the corresponding output.

```
len([1, 2, 3])
3
y = len([1, 2, 3])
y = len([1, 2, 3])
print(y)
3
```

Another note for the Jupyter user is that in Jupyter Notebook, only the **last expression** in a cell is displayed automatically.

Statements require print() to display output.

Here is an example shown in Table 3.5.

TABLE 3.5 Different ways to get the output with and without the print command.

```
10
20
30
30
print(10)
print(20)
print(30)
10
20
30
```

## Role of different IDEs in displaying output

At this stage, it is important to understand that Python executes code in the same logical manner in all environments. The rules governing expressions, statements, assignments, and function calls remain identical whether the user works in Jupyter Notebook, IDLE, VS Code, PyCharm, or the command prompt. The internal working of Python does not change.

However, what differs from one environment to another is how the output is displayed.

Interactive environments such as **Jupyter Notebook** and **Python Shell (IDLE)** are designed mainly for learning, experimentation, and quick testing of commands. In such environments, Python automatically displays the result of expressions (in Jupyter, specifically the last expression in a cell). This helps students immediately see results without writing additional commands.

On the other hand, when Python code is written and executed as a script file (.py) in IDEs such as VS Code or PyCharm, the program is treated as a complete unit. In this case, Python executes all statements, but it does not automatically display intermediate results. The programmer must explicitly use the print() function to display output. This behavior is intentional because script files are meant for building structured programs rather than testing individual expressions.

Therefore, if a command produces no visible output, it does not mean Python is not working. It simply means the environment is waiting for explicit instructions to display the result.

Understanding this distinction prevents confusion when switching between Jupyter Notebook and script-based programming and helps the learner develop clarity about how Python execution and output display are related but conceptually different processes.

## Defining lists in Python

In Python, a list is an ordered collection of elements enclosed in square brackets [].
Lists can store numbers, strings, or mixed data types. In scientific computing, lists are commonly used to store numerical data such as observations, coefficients, or measured values.

A list is defined by assigning values to a variable name.

**Example:**

r = [11, 62, 35, 64, 50]

Unlike MATLAB, Python does not differentiate between row and column vectors at the list level. A list is simply a one-dimensional sequence.

### Generating sequences using range()

To generate sequences with a constant difference, Python provides the range() function.

The syntax is:

```
range(start, stop, step)
```

- start → initial value
- stop → terminal value (excluded)
- step → common difference

By default, the step value is +1, and it does not need to be specified explicitly. Since range() returns a range object, it is typically converted into a list using list(). Some more examples of vector creation in Python are presented in Table 3.6.

TABLE 3.6 Examples of vector creation in Python.

| Requirement | Command | Output |
|---|---|---|
| Define a list with elements from 2 to 13 | list(range(2, 14)) | [2, 3, 4, 5, 6, 7, 8, 9, 10, 11, 12, 13] |
| Define a list with even elements from 1 to 20 | list(range(2, 21, 2)) | [2, 4, 6, 8, 10, 12, 14, 16, 18, 20] |
| Define a list with odd elements from 1 to 20 | list(range(1, 20, 2)) | [1, 3, 5, 7, 9, 11, 13, 15, 17, 19] |
| Define a list with multiples of 3 from 1 to 20 | list(range(3, 21, 3)) | [3, 6, 9, 12, 15, 18] |
| Define a list in reverse order from 20 to 1 | list(range(20, 0, −1)) | [20, 19, 18, ..., 1] |
| Define a list in reverse order from 20 to 1 with step 3 | list(range(20, 0, −3)) | [20, 17, 14, 11, 8, 5, 2] |

### Basic operations on Python lists

After defining a list, it is important to understand how to manipulate and modify its elements. Python provides several built-in methods that allow users to efficiently work with lists. These operations are essential for handling data in scientific and computational tasks. To understand the operations, consider the list: x = [1, 2, 3]. Table 3.7 presents the operations along with their explanations on the list x.

TABLE 3.7 The operations along with their explanations in Python.

| Objective | Command | Output | Explanation |
|---|---|---|---|
| **Adding elements to a list** | x.append(4) print(x) | [1, 2, 3, 4] | The append() method adds an element at the end of the list. |
| **Inserting elements at a specific position** | x.insert(1, 10) print(x) | [1, 10, 2, 3, 4] | The insert(index, value) method inserts an element at a specified position. |
| **Removing elements from a list** | x.remove(2) print(x) | [1, 10, 3, 4] | The remove() method deletes the first occurrence of the specified value. |
| **Removing the last element** | x.pop() print(x) | [1, 10, 3] | The pop() method removes the last element of the list. |
| **Sorting a list** | x.sort() print(x) | [1, 3, 10] | The sort() method arranges elements in ascending order. |

## Generating values using NumPy

While Python lists are useful for storing collections of values, scientific and numerical computations often require more efficient and mathematically expressive data structures. For this purpose, Python commonly uses the **NumPy** library.

NumPy (Numerical Python) provides powerful tools for working with arrays and performing numerical operations efficiently.

### Importing NumPy

Before working with arrays and advanced numerical structures, NumPy must be imported:

```
import numpy as np
```

The alias np is a widely accepted convention in scientific computing.

### Why NumPy is preferred in scientific computing

Although Python lists are flexible and easy to use, they do not support direct element-wise mathematical operations. NumPy arrays, on the other hand:

- Allow vectorized operations
- Are computationally efficient
- Require less memory
- Provide functionality similar to MATLAB vectors

This makes NumPy arrays especially suitable for scientific computations, simulations, numerical analysis, and data processing tasks.

### Python lists versus NumPy arrays

While Python lists and NumPy arrays may appear similar, they are fundamentally different in terms of performance, and functionality. Understanding this difference is essential to choose the appropriate data structure in scientific computing.

The key differences between the Python list and the NumPy array are presented in Table 3.8.

TABLE 3.8 Key differences between the Python list and the NumPy array.

| Feature | Python list | NumPy array |
|---|---|---|
| **Data type** | Can store mixed data types | Stores elements of the same type |
| **Performance** | Slower for numerical computations | Faster due to vectorization |
| **Operations** | Limited mathematical operations | Supports element-wise operations |
| **Memory usage** | Higher | More efficient |
| **Usage** | General-purpose programming | Scientific and numerical computing |

Examples with output to present key differences between the Python list and the NumPy array are presented in Table 3.9.

It is important to note here that NumPy arrays are specifically designed for numerical computations, whereas Python lists are general-purpose data structures. To decide between Python lists and NumPy arrays, the choice depends on the nature of the task. Lists are suitable for storing mixed data types, working with small datasets, and performing general-purpose programming tasks. Whereas, NumPy arrays are preferred while performing mathematical operations, handling large datasets, and implementing scientific or engineering computations.

TABLE 3.9 Examples with output presenting key differences between the Python list and the NumPy array.

| Python list | NumPy array |
|---|---|
| x = [1, 2, 3]<br>y = [4, 5, 6]<br>print(x + y) | import numpy as np<br>x = np.array([1, 2, 3])<br>y = np.array([4, 5, 6])<br>print(x + y) |
| **Output:** | **Output:** |
| [1, 2, 3, 4, 5, 6] | [5 7 9] |
| **Explanation:** | **Explanation:** |
| Lists are concatenated, not added element-wise. | NumPy performs element-wise addition. |

## Advanced vector and array operations using NumPy

After understanding how NumPy arrays support element-wise operations similar to MATLAB vectors, we now explore more advanced vector and matrix manipulations. These operations form the foundation for scientific computing, numerical methods, and applied mathematics using Python.

### Defining vectors in NumPy

A NumPy vector is created using the array() function.

#### Row vector

```
r = np.array([11, 62, 35, 64, 50])
```

#### Column vector

```
c = np.array([41, 72, 93, 42, 15]).reshape(-1, 1)
```

Unlike MATLAB:

- NumPy uses **zero-based indexing** (index starts from 0).
- There is no inherent distinction between row and column vectors.
- A column vector must be explicitly reshaped.

This structural difference is important when performing matrix multiplication.

### Creating vectors using arange()

The function arange() works similarly to MATLAB's colon operator. Some examples along with their output is presented in Table 3.10.

TABLE 3.10 Creation of vectors using the arrange command.

```
x = np.arange(2, 14)
print(x)
[2 3 4 5 6 7 8 9 10 11 12 13]
a = np.arange(2, 21, 2)
print(a)
[2 4 6 8 10 12 14 16 18 20]
b = np.arange(1, 20, 2)
print(b)
[1 3 5 7 9 11 13 15 17 19]
c = np.arange(20, 0, -1)
print(c)
[20 19 18 17 16 15 14 13 12 11 10 9 8 7 6 5 4 3 2 1]
```

**Important points:**
- The stop value is excluded.
- The step value can be positive or negative.
- Works for both integer and floating-point steps.

### Generating equally spaced values

To generate uniformly spaced values between two numbers, use the command linspace as shown below:
**Input:** np.linspace(2, 13, 6)
**Output:** array([ 2., 4.2, 6.4, 8.6, 10.8, 13.])
**Input:** np.linspace(0, 1, 11)
**Output:** array([0., 0.1, 0.2, 0.3, 0.4, 0.5, 0.6, 0.7, 0.8, 0.9, 1.])
Unlike arange(), linspace() includes both endpoints, divides the interval into equal parts, and is commonly used in plotting and numerical analysis.

### Accessing vector elements

NumPy supports flexible and powerful indexing mechanisms for accessing and modifying vector elements. In Python, indexing begins at 0, meaning the first element of a vector is located at index 0, the second at index 1, and so on. This is known as **zero-based indexing.**

To access elements of a vector, square brackets [] are used immediately after the variable name. The number inside the brackets specifies the position (index) of the desired element.

Understanding indexing is essential because almost all numerical computations involve selecting, modifying, or analyzing specific portions of data. Efficient indexing allows the extracting of sub-vectors, selecting data ranges, performing conditional operations and preparing datasets for numerical algorithms

In scientific computing, the ability to manipulate specific sections of arrays without loops significantly improves performance and clarity.

### Zero-based indexing

Unlike MATLAB, where indexing begins at 1, Python begins indexing at 0.

For example, consider:

X = np.array([10, 20, 30, 40, 50])

The positions are presented in Table 3.11 to give the index per element.

TABLE 3.11 Position of elements as per the index in the considered example.

| Element | Index |
|---|---|
| 10 | 0 |
| 20 | 1 |
| 30 | 2 |
| 40 | 3 |
| 50 | 4 |

This difference is extremely important when converting MATLAB programs into Python.

### Positive and negative indexing

Python allows both forward and backward indexing.

– Positive indexing starts from the beginning.

– Negative indexing starts from the end.

For example:

X[0] # First element

X[-1] # Last element

X[-2] # Second last element

Negative indexing is very useful when the size of the vector is unknown.

### Indexing versus slicing

There are two major ways to access data. Indexing means accessing a single element of a vector, while the **slicing refers to** access a range of elements.

In slicing, the starting index is included, but the ending index is excluded.

For example, X[1:4] includes elements at positions 1, 2, and 3 – but not 4. Python also allows skipping elements using a step value in slicing.

For example, X[0:5:2] means start at index 0, stop before index 5, and move in steps of 2.

This concept is extremely useful in numerical simulations and signal processing.

The examples to get more insights in indexing and slicing are shown in Table 3.12.

TABLE 3.12 Examples of indexing and slicing a vector.

```
X = np.arange(1, 20, 2)
print(X)
[ 1 3 5 7 9 11 13 15 17 19]
X[3] # Third element
np.int64(7)
X[-1] # Last element
np.int64(19)
X[0:4] # First four elements
array([1, 3, 5, 7])
X[-4:] # Last four elements
array([13, 15, 17, 19])
X[[1,3,6]] # Specific positions
array([ 3, 7, 13])
X[1::2] # Elements at even index positions
array([ 3, 7, 11, 15, 19])
```

**Understanding the slicing expression**

X[1::2]

The expression X[1::2] uses Python's general slicing format:

```
array[start: stop: step]
```

Here, **start** refers to index where slicing begins (included), **stop** refers to index where slicing ends (excluded), **step** refers to interval between selected elements.

In the command X[1::2], start = 1, stop = not specified (therefore continues till the end), step = 2.

This means start from index 1 and select every second element until the end of the vector.

**Note on NumPy data types**

When accessing a single element of a NumPy array, the output may appear as np.int64(7) instead of simply 7. This occurs because NumPy stores numerical values using its own fixed-size data types (such as int64 and float64) for computational efficiency and memory management.

Although the display format looks different from a standard Python integer, it behaves exactly like a normal number in arithmetic operations. Therefore, this difference is representational and does not affect computations.

## Mathematical operations within a vector

After understanding how to define vectors and access their elements, the next important step is to perform mathematical operations on them. In scientific and numerical computations, vectors are primarily used for performing calculations such as sum, average, multiplication, and transformation of elements.

NumPy provides built-in functions that allow efficient and concise implementation of these operations. Unlike traditional programming approaches that require explicit loops, NumPy performs computations in a vectorized manner, which significantly improves readability, execution speed, and memory efficiency.

In this section, the mathematical operations that can be performed on a single vector are discussed.

Table 3.13 shows some basic operations.

import numpy as np

V=np.array([11, 23, 54, 1, 10])

TABLE 3.13 Basic operations along with the output.

| Requirement | Command | Output | Explanation |
|---|---|---|---|
| Find the number of elements in the vector | len(V) | 5 | Returns the total number of elements. This is a Python built-in function. |
| Find the dimension of the vector | V.shape | (5,) | Returns the shape (dimension) of the array. Since it is one-dimensional, the output is (5,). |
| Compute the sum of elements | np.sum(V) | 99 | Adds all elements of the vector. (11 + 23 + 54 + 1 + 10 = 99). |
| Compute the mean (average) | np.mean(V) | 19.8 | Calculates the arithmetic average of all elements. |
| Compute the product of elements | np.prod(V) | 136620 | Multiplies all elements together. |
| Reverse the order of elements | np.flip(V) | array([10, 1, 54, 23, 11]) | Reverses the vector without modifying the original unless reassigned. |
| Append the elements to the vector | V = np.append(V, [10, 20, 40]) V | array([11, 23, 54, 1, 10, 10, 20, 40]) | Result in a vector V with 3 elements appended at the end. |

## Element-wise operations between two vectors

The above-defined set of commands is all applied on a defined vector. There are many other commands that operate on the elements of two or more vectors and providing meaningful output. Here the two vectors S and R are considered with just three elements each:

```
import numpy as npy
S = np.arange(1, 13, 3)
R = np.arange(2, 10, 2)
```

The discussion of the operations between the two defined vectors S and R is shown in Table 3.14. Be sure to take care of the dimension as it needs to be similar for the operations that require element wise operations.

TABLE 3.14 Operations between vectors S and R.

```
S
array([ 1, 4, 7, 10])
R
array([2, 4, 6, 8])
S+R
array([ 3, 8, 13, 18])
S - R
array([-1, 0, 1, 2])
S * R
array([ 2, 16, 42, 80])
2 ** R
array([ 4, 16, 64, 256])
S @ R
np.int64(140)
```

## Element-wise computation versus matrix multiplication

In computations science, it is important to learn the difference between the element- wise operations and matrix multiplication as per the operations required. Following are the details of the element wise and matrix multiplication with example.

### Element-wise operations

Element-wise operations are performed between corresponding elements of two arrays of the same size. If S and R are the two vectors then S * R multiplies each element of vector S with the corresponding element of vector R.

**Example:**

```
S = np.array([1, 4, 7, 10])
R = np.array([2, 4, 6, 8])
print(S * R)
```

**Output:** [ 2 16 42 80]

**Matrix multiplication**

Matrix multiplication follows the rules of linear algebra and is performed using the @ operator in NumPy. To compute the dot product of the two vectors S and R, the command is S @ R.

**Example:** print(S @ R)
**Output:** 140
Matrix multiplication can also be performed using np.dot(S, R).
Table 3.15 details the difference between the MATLAB and Python operators for the element by element and the matrix multiplication.

TABLE 3.15 Element-by-element operation in MATLAB and Python.

| | MATLAB | Python (NumPy) |
|---|---|---|
| **Element-wise multiplication** | S .* R | S * R |
| **Matrix multiplication** | S * R | S @ R |

### Working with arrays (matrices)

NumPy supports two-dimensional (2D) arrays referred to as matrices, which are extensively used in linear algebra, numerical methods, engineering computations, and scientific modeling. While vectors represent one-dimensional data, matrices represent structured data arranged in rows and columns.

In NumPy, a matrix is simply a 2D array. It can be defined in different ways depending on whether the elements are specified manually, generated using sequences, or created using built-in initialization functions.

The matrix can generally be defined in three different ways:

1. By directly specifying elements using nested lists
2. By generating rows or columns programmatically (e.g., using arange() or linspace())
3. By using built-in matrix generator functions

The general syntax to define a 2D NumPy array is:

```
A=np.array([[row1_elements],
            [row2_elements],
            [row3_elements]])
```

Each inner list represents a row of the matrix. More details with examples are presented in Table 3.16.

Three different examples are discussed above to generate matrix as A, B, and C. The matrix C has the first row: generated using np.arange(); the second row: generated using np.linspace() and the third row is defined manually.

Note that all the rows must have the same number of elements. Otherwise, NumPy will not create a proper 2D array.

TABLE 3.16 Different ways to generate matrix with examples.

```
import numpy as np

A=np.array([[1, 3, 6],
        [11, 0, 9],
        [3, 5, 12]])

print(A)
[[ 1 3 6]
[11 0 9]
[ 3 5 12]]

B=np.array([[2, 4, 6, 8],
[1, 3, 5, 7]])

print(B)
[[2 4 6 8]
[1 3 5 7]]

C=np.array([np.arange(1, 5),
np.linspace(10, 40, 4),
[7, 8, 9, 10]])

print(C)
[[ 1. 2. 3. 4.]
[10. 20. 30. 40.]
[ 7. 8. 9. 10.]]
```

### Built-in matrix generators

While doing the computational, several times there arise a need for use the inbuilt matrix. The frequently used matrices for initialization in numerical algorithms are given with examples in Table 3.17.

TABLE 3.17 Examples to understand the working of matrix generators.

```
import numpy as np

x=np.zeros((3, 4))
print(x)
[[0. 0. 0. 0.]
[0. 0. 0. 0.]
[0. 0. 0. 0.]]
y=np.ones((5, 6))
print(y)
[[1. 1. 1. 1. 1. 1.]
[1. 1. 1. 1. 1. 1.]
[1. 1. 1. 1. 1. 1.]
[1. 1. 1. 1. 1. 1.]
[1. 1. 1. 1. 1. 1.]]

w=np.eye(3)
print(w)
[[1. 0. 0.]
[0. 1. 0.]
[0. 0. 1.]]

z=np.random.rand(4, 4)
print(z)
[[0.49347927 0.2569188 0.16291151 0.72620194]
[0.90518453 0.92587393 0.70736972 0.54510582]
[0.54757256 0.58211658 0.65723777 0.71948922]
[0.71857446 0.98654938 0.59882471 0.2219437]]
```

### Accessing matrix elements

Consider the matrix A, created using the following set of command with details of accessing matrix elements in Table 3.18:

```
import numpy as np
A=np.array([[10, 20, 30], [40, 50, 60], [70, 80, 90]])
print(A)
[[10 20 30]
 [40 50 60]
 [70 80 90]]
```

TABLE 3.18 Examples of accessing matrix elements.

```
A[1, 0] # Element in second row, first column
np.int64(40)
A[:, 1] # Entire second column
array([20, 50, 80])
A[1,:] # Entire second row
array([40, 50, 60])
A[0:2, 1:3] # Submatrix
array([[20, 30],
    [50, 60]])
```

**Important note**
Python uses zero-based indexing. Hence, first row has index 0, while second row has ind ex 1. Table 3.19 differentiates between the MATLAB and Python commands in detail.

TABLE 3.19 Examples of extraction in MATLAB and Python.

| Operation | MATLAB syntax | Python (NumPy) syntax | Description |
|---|---|---|---|
| Define matrix | A=[10 20 30; 40 50 60; 70 80 90] | A=np.array([[10,20,30],[40,50,60],[70,80,90]]) | Create a 3×3 matrix |
| First element | A(1,1) | A[0,0] | MATLAB starts at 1, Python at 0 |
| Second row, first column | A(2,1) | A[1,0] | Access specific element |
| Entire column | A(:,2) | A[:,1] | All rows of a column |
| Entire row | A(2,:) | A[1,:] | All columns of a row |
| Submatrix | A(1:2,2:3) | A[0:2,1:3] | Range extraction |
| Last element | A(end,end) | A[-1,-1] | Last row, last column |
| Last row | A(end,:) | A[-1,:] | Entire last row |
| Last column | A(:,end) | A[:,-1] | Entire last column |

### Deleting rows and columns

Let us proceed to the concept of deleting the rows and column of a given matrix. Consider the above-considered matrix A as an example to learn the syntax and check the output. The matrix A is defined as:

```
import numpy as np
A=np.array([[10, 20, 30], [40, 50, 60], [70, 80, 90]])
print(A)
```

To delete a row or column, the function is deleted with syntax:

```
np.delete(array, index, axis=0 or axis=1)
```

The parameter axis=0 refers to rows, and axis=1 refers to columns. See examples in Table 3.20.

TABLE 3.20 Examples of deleting rows and column from the matrix.

```
B=np.delete(A, 0, axis=0)
print(B)                    # Delete first row

[[40 50 60]
[70 80 90]]

B=np.delete(A, 1, axis=0)
print(B)                    # Delete second row

[[10 20 30]
[70 80 90]]

C=np.delete(A, 0, axis=1)
print(C)                    # Delete first column

[[20 30]
[50 60]
[80 90]]

C=np.delete(A, 1, axis=1)
print(C)                    # Delete second column
[[10 30]
[40 60]
[70 90]]
```

### Deleting multiple rows or columns

One can delete multiple indices at once by passing a list.

**Example:** Delete first and third rows:
D=np.delete(A, [0, 2], axis=0)
print(D)
**Output:** [[40 50 60]]

**Example:** Delete first and third columns:
E=np.delete(A, [0, 2], axis=1)
print(E)
**Output:**
[[20]
[50]
[80]]

### Comparison with MATLAB

**MATLAB:**
A(1,:)=[] → delete first row
A(:,1)=[] → delete first column

**Python (NumPy):**
np.delete(A, 0, axis=0) → delete first row
np.delete(A, 0, axis=1) → delete first column

### Matrix manipulation

While working with the matrices, it is required to get the size and dimension of the considered matrices. NumPy provides several functions to analyze the structure of an array. These functions play an important role in scientific computing, especially when performing matrix operations, reshaping data, and ensuring compatibility between arrays.

**Key properties of a matrix**

- **shape** gives the number of rows and columns of the matrix.
- **size** gives the total number of elements in the matrix.
- **len()** gives the length of the first dimension (i.e., number of rows).

Consider matrix B for demonstrating the commands and description in Table 3.21 in detail:

```
import numpy as np
B=np.array([[1, 3, 9], [2, 4, 6], [5, 1, 7], [10, 8, 15]])
print(B)
```

```
[[ 1 3 9]
[ 2 4 6]
[ 5 1 7]
[10 8 15]]
```

TABLE 3.21 Matrix manipulation in NumPy.

| Command | Description | Output |
|---|---|---|
| **B.shape** | Dimensions of the matrix (rows, columns) | (4, 3) |
| **max(B.shape)** | Largest dimension | 4 |
| **B.size** | Total number of elements | 12 |
| **len(B)** | Number of rows (first dimension) | 4 |

### Reshaping arrays in NumPy

In scientific computing, it is often necessary to change the shape or dimensions of an array without changing the data. NumPy provides the reshape() function to reorganize the elements of an array into a different dimension. The total number of elements must remain the same during reshaping. It does not make any change in the data; it only changes organization of the data.

Here is the syntax to reshape a matrix:

```
array.reshape(rows, columns)
```

**Example:**

```
import numpy as np
x=np.arange(1, 13)
print(x)
```

**Output:**

```
[ 1 2 3 4 5 6 7 8 9 10 11 12]
```

To reshape the matrix x into a 3 × 4 matrix:

A=x.reshape(3, 4)
print(A)

**Output:**
[[ 1 2 3 4]
[ 5 6 7 8]
[ 9 10 11 12]]

**Reshaping into Column Vector**
x.reshape(-1, 1)
Output:
[[ 1]
[ 2]
[ 3]
...
[12]]
Explanation:
– -1 allows NumPy to automatically determine the required dimension.

### Reshaping into row vector

x.reshape(1, -1)
Output:
[[ 1 2 3 4 5 6 7 8 9 10 11 12]]
Here, the row is considered as 1, but for column, there is no specific choice, hence provided as -1.

### Concatenation of arrays

In addition to reshaping, NumPy also allows combining multiple arrays into a single array known as concatenation and is commonly used when working with datasets and matrices in scientific computing.

For example, two arrays A and B can be joined vertically using np.concatenate ((A, B), axis=0) and horizontally using np.concatenate((A, B), axis=1). For vertical concatenation, the number of columns must be the same, while for horizontal concatenation, the number of rows must match.

This concept is useful in combining data from different sources and will be explored in more detail in later chapters.

### Mathematical operations on matrices in NumPy

NumPy provides a wide range of built-in functions to perform mathematical operations on matrices that can be used for the statistical and mathematical operations.

To demonstrate these operations, consider the matrix $W$:

```
import numpy as np
    W=np.array([[2, 3, 0, 8], [34, 12, 11, 4], [41, 27, 13, 7], [17, 13, 19, 9]])
    print(W)
```

**Output:**
[[ 2 3 0 8]
[34 12 11 4]
[41 27 13 7]
[17 13 19 9]]
The commonly used mathematical operations along with the description of command and the output are provided in Table 3.22.

TABLE 3.22 Examples to perform mathematical operations.

| Requirement | Command | Output | Explanation |
|---|---|---|---|
| **Sum of all elements** | np.sum(W) | 220 | Adds all elements of the matrix |
| **Sum column-wise** | np.sum(W, axis=0) | [94 55 43 28] | Sum of each column |
| **Sum row-wise** | np.sum(W, axis=1) | [13 61 88 58] | Sum of each row |
| **Mean of all elements** | np.mean(W) | 13.75 | Average of all elements |
| **Median of all elements** | np.median(W) | 11.5 | Median of all values |
| **Trace of matrix** | np.trace(W) | 36 | Sum of diagonal elements |
| **Determinant** | np.linalg.det (W) | -44,302 | Scalar value representing matrix determinant |
| **Rank of matrix** | np.linalg. matrix_rank(W) | 4 | Number of linearly independent rows/columns |
| **Diagonal elements** | np.diag(W) | [ 2 12 13 9] | Extracts main diagonal |
| **Eigenvalues and eigenvectors** | np.linalg.eig(W) | (eigenvalues, eigenvectors) | Returns eigenvalues and corresponding eigenvectors |
| **Inverse of matrix** | np.linalg.inv(W) | Matrix output | Computes inverse (if matrix is non-singular) |

These functions enable advanced matrix computations required in engineering, physics, and applied mathematics.

### Solving systems of linear equations

One of the most important applications of matrices in scientific computing is to solve the system of linear equations. A system of linear equations can be expressed in matrix form as:

AX=B

where A is the coefficient matrix, X is the vector of unknown variables, and B is the constant (right-hand side) vector

Consider the following system of linear equations:

$x + 2y + 5z = 12$

$-4x + y + 2z = 13$

$9y + 8z = 1$

It can be represented in the matrix form as:

$$\begin{bmatrix} 1 & 2 & 5 \\ -4 & 1 & 2 \\ 0 & 9 & 8 \end{bmatrix} \begin{bmatrix} x \\ y \\ z \end{bmatrix} = \begin{bmatrix} 12 \\ 13 \\ 1 \end{bmatrix}$$

To solve the system of equation, let's define the matrix and proceed:

```
import numpy as np

A = np.array([[1, 2, 5],
          [-4, 1, 2],
          [0, 9, 8]])

B = np.array([12, 13, 1])
```

The solution can be obtained using the function:

```
X = np.linalg.solve(A, B)
print(X)
```

**Output:**

[-2.0317 -3.6984 4.2857]

Another method to solve the system of equation is based on the relation $X=A^{-1}B$ to get the solution. The command to solve the system can be defined as:

```
X=np.linalg.inv(A) @ B
print(X)
```

The function np.linalg.solve() is preferred over np.linalg.inv() because it is more numerically stable and computationally efficient.

**MATLAB versus Python comparison**
In MATLAB, solution can be obtained using: X = A\B, while in Python (NumPy), the command is X = np.linalg.solve(A, B).

## Exercise

### Section A: vector operations

1. Define a = 2.5 and b = 4.2. Compute:
   I. a*b
   II. a**b
   III. b/a
   IV. a+b
2. Create a vector k from 5 to 45 with step 5.
3. Create six equally spaced elements between 10 and 50.
4. Extract positive values from y = sin(t).
5. Double elements greater than 20 using logical indexing.

### Section B: array manipulation

6. Perform horizontal and vertical concatenation.
7. Replace the second column of a random matrix with zeros.
8. Create special structured matrices using zeros() and eye().

### Section C: Advanced structures

9. Solve $AX = B$ using np.linalg.solve().
10. Create tridiagonal and pentadiagonal matrices using np.diag().

## Computational practice questions with short answers

**Example 1: Creating a list of first 10 natural numbers**
To generate the first 10 natural numbers, we use Python's range() function.

**Solution:**
```
L=list(range(1, 11))
```
This creates a list from 1 to 10.

**Output:**
[1, 2, 3, 4, 5, 6, 7, 8, 9, 10]

**Example 2: Sum of elements in a list**
To find the sum of all elements in a list, we use the built-in sum() function.

**Solution:**
```
L=[1, 2, 3, 4]
sum(L)
```

**Output:**
10

**Example 3: Maximum and minimum in a list**
To determine the largest and smallest elements, we use max() and min().

**Solution:**
```
L=[5, 2, 9, 1]
max(L), min(L)
```

**Output:**
9, 1

**Example 4: Reversing a list**
A list can be reversed using slicing.

**Solution:**
```
L=[1, 2, 3]
L[::-1]
```

**Output:**
[3, 2, 1]

**Example 5: Removing duplicates from a list**
Duplicates can be removed by converting the list into a set.

**Solution:**
```
L = [1, 2, 2, 3]
list(set(L))
```

**Output:**

[1, 2, 3]

**Example 6: Generating squares of numbers**

List comprehension is used to generate squares.

**Solution:**

```
[x**2 for x in range(1, 11)]
```

**Output:**

[1, 4, 9, 16, 25, 36, 49, 64, 81, 100]

**Example 7: Extracting even numbers from a list**

Filtering can be done using list comprehension.

**Solution:**

```
L=[1, 2, 3, 4, 5]
[x for x in L if x % 2==0]
```

**Output:**

[2, 4]

**Example 8: Converting temperature (°C to °F)**

Formula used: $F = \frac{9}{5}C + 32$

**Solution:**

```
C = [0, 20, 30]
[(9/5)*c + 32 for c in C]
```

**Output:**

[32.0, 68.0, 86.0]

**Example 9: Normalizing a list**

Each element is divided by the maximum value.

**Solution:**

```
L = [2, 4, 6]
[x / max(L) for x in L]
```

**Output:**

[0.333 ..., 0.666 ..., 1.0]

**Example 10: Computing sine values**
Using the math library to compute sine values.

**Solution:**
```
import math
angles=[0, math.pi/2, math.pi]
[math.sin(a) for a in angles]
```

**Output:**
```
[0.0, 1.0, 0.0]
```

**Example 11: Creating a NumPy array**
NumPy arrays are efficient for numerical operations.

**Solution:**
```
import numpy as np
np.array([1, 2, 3])
```

**Output:**
```
[1 2 3]
```

**Example 12: Adding scalar to array**
A scalar value is added to each element.

**Solution:**
```
arr = np.array([1, 2, 3])
arr + 2
```

**Output:**
```
[3 4 5]
```

**Example 13: Multiplying array by scalar**
Each element is multiplied by a constant.

**Solution:**
```
arr * 3
```

**Output:**
```
[3 6 9]
```

**Example 14: Element-wise Square**
Each element is squared.

**Solution:**
arr**2

**Output:**
[1 4 9]

**Example 15: Square Root of Elements**
Using NumPy's square root function.

**Solution:**
np.sqrt([4, 9, 16])

**Output:**
[2. 3. 4.]

**Example 16: Mean of an array**
The average value is computed using mean().

**Solution:**
np.mean([2, 4, 6, 8])

**Output:**
5.0

**Example 17: Standard deviation**
Measures dispersion in data.

**Solution:**
np.std([2, 4, 6, 8])

**Output:**
2.236 ...

**Example 18: Variance**
Variance is the square of standard deviation.

**Solution:**
np.var([2, 4, 6, 8])

**Output:**

5.0

**Example 19: Adding two arrays**

Element-wise addition of arrays

**Solution:**

```
a = np.array([1, 2])
b = np.array([3, 4])
a + b
```

**Output:**

[4 6]

**Example 20: Dot product**

Computes the scalar product of two vectors

**Solution:**

```
np.dot([1, 2], [3, 4])
```

**Output:**

11

**Example 21: Cumulative sum**

Computes running total of list elements

**Solution:**

```
L = [1, 2, 3, 4]
cumsum = []
s = 0
for x in L:
s + = x
cumsum.append(s)
```

**Output:**

[1, 3, 6, 10]

**Example 22: Creating a 2 × 2 matrix**

A matrix can be created using NumPy.

**Solution:**

```
np.array([[1, 2], [3, 4]])
```

**Output:**

```
[[1 2]
[3 4]]
```

**Example 23: *Z*-score normalization**
Used to standardize data

**Solution:**

```
import math
L = [1, 2, 3]
mean = sum(L)/len(L)
std = math.sqrt(sum((x-mean)**2 for x in L)/len(L))
[(x-mean)/std for x in L]
```

**Output:**

```
[-1.224 ..., 0.0, 1.224 ... ]
```

**Example 24: Weighted average**
Weights are applied to values.

**Solution:**

```
values = [10, 20, 30]
weights = [1, 2, 1]
sum(v*w for v, w in zip(values, weights)) / sum(weights)
```

**Output:**

```
20.0
```

**Example 25: Solving linear system**
Using NumPy to solve $Ax = b$

**Solution:**

```
A = np.array([[1, 2], [3, 4]])
b = np.array([5, 11])
np.linalg.solve(A, b)
```

**Output:**

```
[1. 2.]
```

## Mini project: numerical study of functions using lists and arrays

### Analysis of mathematical functions using Python lists and arrays

#### Objectives

- Represent mathematical functions numerically
- Use lists and arrays for computation
- Apply mathematical concepts like derivatives and integration
- Compare computational approaches

### Problem statement with hints

***Step 1: Data generation***
Generate values of xx in a given interval (e.g., 0–10) and compute $f(x)$.
**Hints**
- Use range() or a loop to generate values
- Store values in a list
- Use list comprehension for cleaner code
- Choose a function like:
  $x^2$, $sin(x)$, or $e^{-x}$
- Use the math module for non-polynomial functions

***Step 2: List-based computations***
Perform operations using only lists.
**Tasks**
- Compute first differences
- Compute moving average (window = 3)
- Find max and min
- Filter values greater than mean

**Hints**
- First difference: subtract consecutive elements → think about indexing (i and i+1)
- Moving average: slice list segments → L[i:i+3]
- Mean = sum/length
- Filtering: use list comprehension with condition

***Step 3: Numerical differentiation***
Approximate derivative using forward difference.
**Hints**
- Use formula: difference in $y$-values divided by difference in $x$-values

- Use consecutive points only
- Result list will have one fewer element than original
- Compare pattern with expected derivative (e.g., slope increasing or constant)

*Step 4: Numerical integration*
Approximate area under curve
**Hints**
- Use trapezoidal rule logic:
  - First and last values get half weight
  - Middle values get full weight
- Multiply by step size
- Use sum() for middle values

*Step 5: Array-based computations*
Repeat steps using arrays
**Hints**
- Import NumPy
- Convert list → array using np.array()
- Use vectorized operations (no loops)
- Differences: use built-in NumPy functions
- Filtering: use boolean conditions (arr > value)

*Step 6: Statistical analysis*
Compute statistical measures
**Hints**
- Mean: average of values
- Variance: measure of spread
- Standard deviation: square root of variance
- Outliers:
  - Think: values far from mean
    - Use condition like: mean $\pm$ k $\times$ standard deviation

*Step 7: Comparison study*
Analyze results
**Hints**
- Compare:
  - Accuracy (numerical versus expected trend)
  - Speed (arrays versus lists – conceptually)
- Think:
  - Why arrays are faster?
  - Where lists are still useful?

CHAPTER 4

# Working with data types

Data types are an essential concept in programming. They play an important role in defining the nature of data. Using a correct data type as per the requirement is essential for the efficient usage of memory and proper handling of the data.

In programming, every value is associated with a data type and thus defining the kind of data being stored is crucial to find the operations that can be performed on it.

## Common built-in data types

Both MATLAB and Python support a variety of data types for storing and manipulating data. While the purpose of these data types is similar in both environments, their representation and usage differ. Understanding these differences is important for users transitioning from MATLAB to Python. Table 4.1 lists out the commands defining the data types in MATLAB in comparison to Python.

TABLE 4.1 Different data types in MATLAB and Python.

| Concept | MATLAB | Python | Description |
|---|---|---|---|
| **Integer** | int8, int16, int32 | int | Whole numbers |
| **Floating point** | double (default) | float | Decimal numbers |
| **Complex numbers** | 3 + 4i | 3 + 4j | Complex values |
| **String/text** | 'Hello' | "Hello" | Character or text data |
| **Boolean** | true, false | True, False | Logical values |
| **Array/vector** | [1 2 3] | [1, 2, 3] (list) / np.array([1,2,3]) | Collection of elements |
| **Matrix** | [1 2; 3 4] | np.array([[1,2],[3,4]]) | 2D data structure |
| **Cell array** | {1, 'a', 3} | List | Mixed data types |
| **Structure** | struct | Dictionary (dict) | Key–value pairs |

In MATLAB, most data are considered and saved as arrays, whereas Python provides a wider variety of data types suitable for general-purpose programming as well as scientific computing.

 | HTTPS://DOI.ORG/10.1515/9783112207758-004

## Variables and naming conventions in Python

A variable is a name used to store data in a program. In Python, variables are created when a value is assigned to them using the assignment operator = . Unlike some programming languages, Python does not require explicit declaration of a variable's data type.

### Creating variables

x = 10
y = 3.14
name = "Python"
In the above examples, variables are automatically assigned data types based on the values they store.

### Rules for naming variables

When defining variable names in Python, the following rules must be followed:
- A variable name must start with a letter (a–z, A–Z) or an underscore _.
- It cannot start with a number.
- It can contain letters, digits, and underscores.
- Variable names are case-sensitive (x and X are different).
- Reserved keywords (such as if, for, while) cannot be used as variable names.

#### Naming conventions

Although Python allows flexibility in naming variables, certain conventions are recommended for better readability and consistency.
1. **Snake case (recommended in Python)**
total_marks = 85
student_name = "Ravi"
This is the most commonly used convention in Python in which the variable names are separated using underscores.
2. **Camel case**
totalMarks = 85
studentName = "Ravi"
In this case, the first word starts with lowercase and the subsequent words start with uppercase.
3. **Pascal case**
TotalMarks = 85
StudentName = "Ravi"

In Pascal case, each word starts with an uppercase letter and is often used for class names.

### Indentation in Python

In Python, indentation is used to define the structure of a program. Unlike many other programming languages that use braces { } to define blocks of code, Python relies on indentation (spacing at the beginning of a line) to group statements. Indentation is not just for readability but it is a requirement in Python. Incorrect indentation will result in an error.

**Example: correct indentation**

```
for i in range(3):
print(i)
```

**Output:**

```
0
1
2
```

The print(i) statement is indented, indicating that it belongs to the for loop.

**Example: incorrect indentation**

```
for i in range(3):
print(i)
```

This will produce an error because the statement inside the loop is not properly indented.

To avoid indentation errors, it is recommended to use four spaces consistently. All statements within the same block must have the same indentation level. Mixing tabs and spaces should be avoided.

### The `print()` function in Python

The print() function is used to display output in Python. Unlike MATLAB, where results are displayed automatically (unless suppressed), Python requires the use of print() to explicitly display values when working in script-based environments.

### Basic usage

print("Hello, Python")

**Output:**
Hello, Python

### Printing variables

x = 10
print(x)

**Output:**
10

### Printing multiple variables

Python allows printing multiple values in a single statement:
x = 5
y = 10
z = 15
print(x, y, z)

**Output:**
5 10 15
Values are separated by spaces by default.

### Custom separator

To use a separator other than the default space, the sep parameter can be used as shown below:
print(x, y, z, sep = ", ")

**Output:**
5, 10, 15

### Using + for concatenation

The + operator can be used to concatenate strings. The + operator works for string concatenation only when all elements are strings else can lead to errors. The following example clarifies the difference:

```
name = "Python"
print("Hello " + name)
```

**Output:**
Hello Python

**Incorrect usage:**

```
x = 10
print("Value: " + x)
```

This produces an error.

**Correct usage:**

```
print("Value:", x)
OR
print("Value: " + str(x))
```

## Type conversion in Python

It is often required to convert data from one type to another known as **type conversion** or **type casting**. It is useful while performing mathematical operations, handling user input, or combining different types of data. Details are provided in Table 4.2.

TABLE 4.2 Common type conversion functions.

| Function | Description |
|---|---|
| **int()** | Converts a value to an integer |
| **float()** | Converts a value to a floating-point number |
| **str()** | Converts a value to a string |
| **bool()** | Converts a value to a boolean |

The examples to convert the given data from one defined type to other as per requirement are presented in Table 4.3.

It is important to note that converting a float to an integer removes the decimal part without rounding. String and numeric types must be converted before combining using the + operator. Type conversion ensures compatibility between different data types and helps prevent errors during computations.

TABLE 4.3 Commands of conversion along with the output.

| Purpose | Commands | Output |
|---|---|---|
| **Converting to integer** | x = 5.7<br>y = int(x)<br>print(y) | 5 |
| **Converting to float** | x = 10<br>y = float(x)<br>print(y) | 10.0 |
| **Converting to string** | x = 25<br>y = str(x)<br>print("Value is " + y) | Value is 25 |
| **Converting to boolean** | print(bool(0))<br>print(bool(5)) | False<br>True |

## Functions in Python

A function is a block of organized and reusable code that performs a specific task. Functions help in breaking a program into smaller, manageable parts, making the code more structured, readable, and reusable.

Python provides both **built-in functions** and **user-defined functions.**

### Built-in functions

Python includes several predefined functions that can be used directly. For example, calculating the length, maximum value or the sum of the elements for a vector can be done using the built-in functions.

Consider a vector x = [1, 2, 3, 4] with the built-in functions as follows:

```
print(len(x))      # length of list
print(max(x))      # maximum value
print(sum(x))      # sum of elements
```

Output:

```
4
4
10
```

### Defining a function

In Python, a function can be defined using the keyword def.

**Syntax**

```
def function_name(parameters):
statements
return value
```

**Example: calculating square of a number**

```
def square(x):
return x * x

print(square(5))
```

**Output:**

25

### Function with multiple parameters

The function can be defined on one or more input values and providing the corresponding required output. Here is an example of a function that takes two numbers to be added and return the sum.

```
def add(a, b):
    return a + b
print(add(3, 7))
Output:
10
```

### Function without return statement

For the function usage, there is no restriction on returning the output values, a function may have an output after performing a task or may have no specific output but is performing a specific task in that call. Following is an example of function in which the function is used to print a string with Hello as a message:

```
def greet(name):
    print("Hello", name)
greet("Python")
```

Output:
Hello Python

## Taking user input in Python

In many programs, it is necessary to accept input from the user during execution. Python provides the input() function to read data entered by the user through the keyboard.

**Basic syntax**

```
variable=input("message")
```

Here, the message inside quotes is displayed to the user as a string with the entered value stored in the variable.

**Example:**
name = input("Enter your name: ")
print("Hello", name)

**Output (sample run):**
Enter your name: Ravi
Hello Ravi

**Converting input to numeric type**
To perform numerical operations, the input must be converted using type conversion functions:
x = int(input("Enter a number: "))
y = float(input("Enter another number: "))
print(x + y)

**Example output:**
Enter a number: 5
Enter another number: 2.5
7.5

**Common mistake:**
x = input("Enter number: ")
print(x + 5)
This produces an error because x is treated as a string.

## Variables and naming conventions in MATLAB (comparison with Python)

In the above section, the details are provided for Python. This section is to detail all the conventions and command for MATLAB in comparison to Python.

In MATLAB, variables are also used to store data values. Similar to Python, variables are created using the assignment operator = . However, MATLAB treats most variables as arrays by default, even when storing a single value.

**Defining variables**

In MATLAB, variables are usually treated as arrays, whereas in Python, variables can represent different data types such as integers, strings, or lists. MATLAB statements typically end with a semicolon (;) to suppress output, whereas Python does not require semicolons.

### Rules for naming variables

The rules for naming variables in MATLAB are similar to Python:
- Must begin with a letter
- Can contain letters, digits, and underscores
- Cannot contain spaces or special characters
- MATLAB is case-sensitive (x and X are different)

**Example:**

total_marks = 85;
studentName = 'Ravi';

Unlike Python, MATLAB does not enforce specific naming conventions such as snake_case or camelCase, but meaningful naming is strongly recommended.

### Output display in MATLAB versus Python

One of the fundamental differences between MATLAB and Python lies in how output is displayed. Understanding this distinction helps avoid confusion when transitioning between the two environments.

In MATLAB, the result of a command is displayed automatically unless it is terminated with a semicolon (;), while Python requires the use of print() to display output in scripts. In Jupyter Notebook, Python displays the result of the last expression automatically, which is similar to MATLAB behavior.

**Example for MATLAB:**

x = 10

Output:

x =

10

If a semicolon is used, the output is suppressed:

x = 10;

Output:

(no output displayed)

The difference in the output can be summarized as follows:

**MATLAB versus Python summary**

The discussion of the features along with the details in brief is given in Table 4.4.

TABLE 4.4 Differentiation of the features of MATLAB and Python.

| Feature | MATLAB | Python |
|---|---|---|
| **Default output** | Automatically displayed | Not displayed in scripts |
| **Suppressing output** | Use; | Not applicable |
| **Explicit display** | Not required | Use print() |
| **Interactive mode** | Always shows output | Shows last expression (Jupyter) |

## Type conversion in MATLAB

In MATLAB, type conversion is also required when working with different types of data. However, unlike Python where conversion is done using functions such as int () or float(), MATLAB uses specific conversion functions depending on the desired data type. Furthermore, the details and differences can be summarized in Table 4.5.

TABLE 4.5 Common type conversion functions for MATLAB and Python.

| Purpose | MATLAB | Python |
|---|---|---|
| **Convert to integer** | int32(x) | int(x) |
| **Convert to float** | double(x) | float(x) |
| **Convert to string** | string(x) or num2str(x) | str(x) |
| **Convert to logical** | logical(x) | bool(x) |

**Example: converting to integer:**

```
x = 5.7;
y = int32(x)
```

**Output:**

```
5
```

**Example: converting to double (floating point):**

```
x = 10;
y = double(x)
```

**Output:**

```
10
```

**Example: converting to string:**

```
x = 25;
y = num2str(x)
```

**Output:**

```
'25'
```

**Example: converting to logical:**

```
logical(0)
logical(5)
```

**Output:**

```
0
1
```

## Functions in MATLAB

In MATLAB, functions are used to perform specific tasks, similar to Python. However, the syntax and structure of defining functions differ from Python. MATLAB functions are typically written in separate files or defined at the end of a script.

### Defining a function in MATLAB

A function in MATLAB is defined using the keyword function.

**Syntax**

```
function output=function_name(input)
    statements
end
```

**Example: calculating square of a number:**

```
function y = square(x)
    y = x^2;
end
```

**Calling the function**

```
square(5)
```

**Output:**

25

### Function with multiple inputs

Here is an example of a function with the name add with two inputs as number to be added and giving an output:

```
function z = add(a, b)
    z = a + b;
end
```

### Function with multiple outputs

The number of input to a function can be more than one and needs to be explicitly handled using two variables to store the values as output. Here is an example:

```
function [sum, diff] = operations(a, b)
    sum = a + b;
    diff = a – b;
end
```

### Comparison with Python

MATLAB functions are defined using the keyword 'function' and typically require an explicit output variable, whereas Python functions are defined using 'def' and use the 'return' statement. Furthermore, in MATLAB, the name of the function file must match the function name when defined in a separate file. The comparison is given in Table 4.6.

TABLE 4.6 Summary of MATLAB and Python.

| Feature | MATLAB | Python |
|---|---|---|
| **Function keyword** | function | def |
| **Output** | Defined in function header | Returned using return |
| **Multiple outputs** | Supported directly | Returned as tuple |
| **File structure** | Often separate file | Can be inline |

### Taking the user input in MATLAB (comparison with Python)

In MATLAB, user input is obtained using the input() function, similar to Python. However, MATLAB allows direct numerical input without requiring explicit type conversion. While in Python, input() always returns a string, and type conversion is required for numerical operations, in MATLAB, input() directly evaluates numeric input, so no conversion is needed for numbers.

**Basic syntax**

```
variable=input('message')
```

The message inside quotes is displayed to the user with the entered value stored in the variable.

**Example: numeric input:**
x = input('Enter a number: ')

**Output (sample run):**
Enter a number: 5
Here, x is stored as a numeric value.

**Example: string input:**
To accept string input, MATLAB requires a special argument 's':
name = input('Enter your name: ', 's')

**Output (sample run):**
Enter your name: Ravi
A brief summary of features of MATLAB and Python is presented in Table 4.7.

TABLE 4.7 Summary of features of MATLAB and Python discussing functions.

| Feature | MATLAB | Python |
|---|---|---|
| **Function** | input() | input() |
| **Numeric input** | Directly numeric | Returned as string |
| **String input** | Use 's' option | Default behavior |
| **Type conversion** | Not required (for numbers) | Required |

**Key concept summary**
MATLAB is primarily designed for numerical and matrix-based computations, where most variables are treated as arrays by default. In contrast, Python provides a flexible programming environment with multiple data types and requires explicit handling of output and type conversions. Understanding these differences helps in effectively transitioning between MATLAB and Python for scientific computing.

## Exercise

### Section A: data types and variables

1. Identify the data type of the following values in Python:
   (a) 25
   (b) 3.14
   (c) "Hello"
   (d) True
2. Define variables in Python and MATLAB for storing:
   (a) an integer
   (b) a floating-point number
   (c) a string
3. Write valid and invalid variable names and justify why they are valid or invalid.

### Section B: output and print

4. Write Python statements to display:
   (a) A single variable
   (b) Multiple variables in one line
   (c) Values separated by commas
5. Write MATLAB commands to:
   (a) display output
   (b) suppress output

### Section C: type conversion

6. Convert the following in Python:
   (a) float to integer
   (b) integer to string
   (c) string to float
7. Perform equivalent conversions in MATLAB using:
   (a) int32()
   (b) double()
   (c) num2str()

### Section D: functions

8. Define a function in Python to compute:
   (a) square of a number
   (b) sum of two numbers
9. Define equivalent functions in MATLAB.
10. Write a function that returns two outputs (sum and difference) in:
    (a) Python
    (b) MATLAB

### Section E: user input

11. Write a Python program to:
    - take two numbers as input
    - print their sum
12. Write a MATLAB program to:
    - take a number as input
    - display its square

## Computational practice questions with short answers

**Q1.** Compute 0.1 + 0.2 = = 0.3 in MATLAB and Python. Explain the result.
**Short answer:**
MATLAB:

| | **Answer:** |
|---|---|
| `0.1 + 0.2 = = 0.3` | 0 |

Python:

| | **Answer:** |
|---|---|
| `0.1 + 0.2 = = 0.3` | False |

**Q2.** Write code to safely check whether 0.1 + 0.2 is approximately equal to 0.3.
**Short answer:**
MATLAB:

| | **Answer:** |
|---|---|
| `abs((0.1 + 0.2) - 0.3) < 1e-12` | 1 |

Python:

| | **Answer:** |
|---|---|
| `abs((0.1 + 0.2) - 0.3) < 1e-12` | True |

**Q3.** In MATLAB, store 300 as an unsigned 8-bit integer. What happens?
**Short answer:**
MATLAB:

| | **Answer:** |
|---|---|
| `x = uint8(300)` | 255 |

Q4. In Python using NumPy, store 250 as uint8 and add 10. What is the result?
**Short answer:**

| | |
|---|---|
| import numpy as np | **Answer:** |
| x = np.uint16(250) | 260 |
| y = x + np.uint16(10) | |
| y | |

Q5. Compare the results of 1/3 + 1/3 + 1/3 and 1 using floating-point arithmetic
**Short answer:**
MATLAB:

| | |
|---|---|
| (1/3 + 1/3 + 1/3) = = 1 | **Answer:** |
| | 1 |

Python:

| | |
|---|---|
| (1/3 + 1/3 + 1/3) = = 1 | **Answer:** |
| | True |

Q6. What data type results from adding an integer and a floating-point number?
**Short answer:**
MATLAB:

| | |
|---|---|
| a = int32(5); | **Answer:** |
| b = 2.5; | int32 |
| result = a + b; | |
| class(result) | |

Python:

| | |
|---|---|
| type(5 + 2.5) | **Answer:** |
| | float |

Q7. Create a vector of integers from 1 to 10 and convert it to floating-point values.
**Short answer:**
MATLAB:

| | |
|---|---|
| x = 1:10; | **Answer:** |
| y = double(x); | y = |
| y | 1 2 3 4 5 6 7 8 9 10 |

Python:

| | |
|---|---|
| import numpy as np | **Answer:** |
| x = np.arange(1, 11) | array([ 1., 2., 3., 4., 5., 6., 7., 8., 9., 10.]) |
| y = x.astype(float) | |
| y | |

Q8. From numbers 1 to 20, extract all values divisible by both 2 and 3.
**Short answer:**
MATLAB:

| | |
|---|---|
| x = 1:20; | **Answer:** |
| result = x(mod(x,2) = = 0 & mod(x,3) = = 0); | result = |
| result | 6 12 18 |

Python:

| | |
|---|---|
| import numpy as np | **Answer:** |
| x = np.arange(1, 21) | array([ 6, 12, 18]) |
| result = x[(x % 2 = = 0) & (x % 3 = = 0)] | |
| result | |

**Q9.** Write a program in MATLAB and Python to compute the area of a circle for a given radius $r = 5$. Display the result in the format: Area = < value >

**Short answer:**

MATLAB:

```
r = 5;
area = pi * r^2;
msg = ['Area = ' num2str(area)];
disp(msg);
```

**Answer:**

Area = 78.5398

Python:

```
import math
r = 5
area = math.pi*r**2
msg = "Area = " + str(area)
msg
```

**Answer:**

'Area = 78.53981633974483'

**Q10.** Store a student name, roll number, and marks in one container.

**Short answer:**

MATLAB:

```
student = {'Asha', 12, [85 90 88]};
student
```

**Answer:**

student =

'Asha' [12] [1x3 double]

Python:

```
student = ["Asha", 12, [85, 90, 88]]
student
```

**Answer:**

['Asha', 12, [85, 90, 88]]

**Q11.** Store radius and height of a cylinder, then compute volume.
**Short answer:**
MATLAB:

```
c.radius = 3;
c.height = 10;
volume = pi*c.radius^2*c.height;
volume
```

**Answer:**
volume =
282.7433

Python:

```
import math
c = {"radius": 3, "height": 10}
volume = math.pi * c["radius"]**2 * c["height"]
volume
```

**Answer:**
282.7433388230814

**Q12.** Compute the magnitude of the complex number 3 + 4i.
**Short answer:**
MATLAB:

```
z = 3 + 4i;
abs(z)
```

**Answer:**
5

Python:

```
z = 3 + 4j
abs(z)
```

**Answer:**
5.0

**Q13.** Find the roots of $x^2 + 4x + 13 = 0$.
**Short answer:**
MATLAB:

```
roots([1 4 13])
```

**Answer:**
ans =
-2.0000 + 3.0000i
-2.0000 − 3.0000i

Python:

| | |
|---|---|
| import numpy as np<br>np.roots([1, 4, 13]) | **Answer:**<br>array([− 2. + 3.j, − 2.-3.j]) |

Q14. Create a 2 × 2 matrix and compute its determinant.
**Short answer:**
MATLAB:

| | |
|---|---|
| A = [2 3; 1 4];<br>det(A) | **Answer:**<br>5 |

Python:

| | |
|---|---|
| import numpy as np<br>A = np.array([[2, 3], [1, 4]])<br>np.linalg.det(A) | **Answer:**<br>5.000000000000001 |

Q15. Convert a matrix of integers into floating-point values before division.
**Short answer:**
MATLAB:

| | |
|---|---|
| A = int32([2 4; 6 8]);<br>B = double(A)/3;<br>B | **Answer:**<br>B =<br>0.6667 1.3333<br>2.0000 2.6667 |

Python:

| | |
|---|---|
| import numpy as np<br>A = np.array([[2, 4], [6, 8]], dtype = int)<br>B = A.astype(float)/3<br>B | **Answer:**<br>array([[0.66666667,<br>1.33333333],<br>[2., 2.66666667]]) |

Q16. From numbers 1 to 30, select perfect squares.
**Short answer:**
MATLAB:

```
x = 1:30;
result = x(mod(sqrt(x),1) = = 0);
result
```

**Answer:**
result =
1 4 9 16 25

Python:

```
import numpy as np
x = np.arange(1, 31)
result = x[np.sqrt(x) % 1 = = 0]
result
```

**Answer:**
array([ 1, 4, 9, 16, 25])

Q17. Generate the cubes of all integers from 1 to 10.
**Short answer:**
MATLAB:

```
x = 1:10;
cubes = x.^3;
cubes
```

**Answer:**
cubes =
Columns 1 through 8
1 8 27 64 125 216 343 512
Columns 9 through 10
729 1,000

Python:

```
cubes = [x**3 for x in range(1, 11)]
cubes
```

**Answer:**
[1, 8, 27, 64, 125, 216, 343, 512, 729, 1,000]

Q18. In Python, why does t[0] = 10 fail when t = (1,2,3)?
**Short answer:**
Python:

```
t = [1, 2, 3]
t[0] = 10
t
```

**Answer:**
[10, 2, 3]

**Q19.** Store marks of three students and compute their average marks.

**Short answer:**

MATLAB:

```
T = table(['A';'B';'C'], [80;90;85], [70;88;92], ...
'VariableNames', {'Name','Math','Science'});
T.Average = (T.Math + T.Science)/2;
T
```

**Answer:**

T =

Name Math Science Average

____ ____ _______ _______

A 80 70 75

B 90 88 89

C 85 92 88.5

Python:

```
import pandas as pd
df = pd.DataFrame({
"Name": ["A", "B", "C"],
"Math": [80, 90, 85],
"Science": [70, 88, 92]
})
df["Average"] = (df["Math"] + df["Science"])
/ 2
df
```

**Answer:**

| Name | Math | Science | Average | |
|---|---|---|---|---|
| 0 | A | 80 | 70 | 75.0 |
| 1 | B | 90 | 88 | 89.0 |
| 2 | C | 85 | 92 | 88.5 |

**Q20.** Write code that checks whether a variable is numeric before computing its square.

**Short answer:**

MATLAB:

```
x = 6;
if isnumeric(x)
y = x^2;
end
y
```

**Answer:**

36

Python:

```
x = 6
if isinstance(x, (int, float)):
y = x**2
y
```

**Answer:**

36

**Q21.** Why does "10" + 5 fail in Python? Fix it.
**Short answer:**
Python:

| | |
|---|---|
| result = int("10") + 5 | **Answer:** |
| result | 15 |

**Q22.** Define $x$ symbolically and expand $(x + 2)^3$.
**Short answer:**
MATLAB:

| | |
|---|---|
| syms x | **Answer:** |
| expand((x + 2)^3) | **X^3 + 6*x^2 + 12*x + 8** |

Python:

| | |
|---|---|
| import sympy as sp | **Answer:** |
| x = sp.symbols('x') | x**3 + 6*x**2 + 12*x + 8 |
| sp.expand((x + 2)**3) | |
| print(sp.expand((x + 2)**3)) | |

**Q23.** Compare numeric and symbolic computation of $\sqrt{2}$.
**Short answer:**
MATLAB:

| | |
|---|---|
| sqrt(2) | **Answer:** |
| syms x | 2^(1/2) |
| sqrt(sym(2)) | |

Python:

| | |
|---|---|
| import math, sympy as sp | **Answer:** $\sqrt{2}$ |
| math.sqrt(2) | |
| sp.sqrt(2) | |

**Q24.** Multiply two matrices correctly in MATLAB and Python.
**Short answer:**
MATLAB:

```
A = [1 2; 3 4];
B = [2 0; 1 2];
C = A * B;
C
```

**Answer:**
C =
4 4
10 8

Python:

```
import numpy as np
A = np.array([[1, 2], [3, 4]])
B = np.array([[2, 0], [1, 2]])
C = A @ B
C
```

**Answer:**
array([[ 4, 4],
[10, 8]])

**Q25.** Calculate the mean of [1, 2, 3, 4, 100] as integer data and floating-point data. Why is float preferred?
**Short answer:**
MATLAB:

```
x = [1 2 3 4 100];
mean(x)
```

**Answer:**
22

Python:

```
import numpy as np
x = np.array([1, 2, 3, 4, 100], dtype = float)
np.mean(x)
```

**Answer:**
22.0

## Mini project: data type-aware student performance analyzer

### Project goal

Build a small program that stores student marks, validates data types, computes mathematical results, and generates a grade report.

### Problem statement

Create a program in **MATLAB and Python** that manages the marks of five students in three subjects: mathematics, physics, and computer science.

The program should:

1. Store student names as text.
2. Store roll numbers as integers.
3. Store marks as numeric arrays/matrices.
4. Compute total marks for each student.
5. Compute percentage for each student.
6. Assign grades using logical conditions.
7. Identify the topper.
8. Find subject-wise average marks.
9. Detect invalid marks below 0 or above 100.
10. Display a clean final report.

**Sample data**

```
Names: Asha, Ravi, Neha, Karan, Meera
Roll Numbers: 101, 102, 103, 104, 105
Marks:
Asha 88 76 92
Ravi 67 81 74
Neha 95 89 91
Karan 45 52 49
Meera 78 85 80
```

### Expected computations

**Each student has three subjects, so:**

```
Total Marks = Math + Physics + Computer
Percentage = Total Marks / 3
```

### Example for Asha:

Total = 88 + 76 + 92 = 256
Percentage = 85.33

#### Grade rules

Percentage > =90 Grade A +
Percentage > =80 Grade A
Percentage >= 70 Grade B
Percentage > =60 Grade C
Percentage < 60 Grade D

#### Hints

### Hint 1: Use appropriate data types MATLAB

```
names = {'Asha','Ravi','Neha','Karan','Meera'};
roll = int32([101 102 103 104 105]);
marks = [88 76 92;
67 81 74;
95 89 91;
45 52 49;
78 85 80];
```

### Python

```
import numpy as np
names = ["Asha", "Ravi", "Neha", "Karan", "Meera"]
roll = np.array([101, 102, 103, 104, 105], dtype = np.int32)
marks = np.array([
[88, 76, 92],
[67, 81, 74],
[95, 89, 91],
[45, 52, 49],
[78, 85, 80]
], dtype = float)
```

### Hint 2: Validate marks
### MATLAB

```
invalid = marks < 0 | marks > 100;
```

### Python

```
invalid = (marks < 0) | (marks > 100)
```

### Hint 3: Compute totals and percentages
### MATLAB

```
total = sum(marks, 2);
percentage = total / 3;
```

### Python

```
total = np.sum(marks, axis = 1)
percentage = total / 3
```

### Hint 4: Find topper
### MATLAB

```
[maxPercent, index] = max(percentage);
topperName = names{index};
```

### Python

```
index = np.argmax(percentage)
topper_name = names[index]
max_percent = percentage[index]
```

### Hint 5: Subject-wise averages

#### MATLAB

```
subjectAverage = mean(marks, 1);
```

#### Python

```
subject_average = np.mean(marks, axis = 0)
```

### Hint 6: Assign grades

#### MATLAB

```
grades = cell(5,1);
for i = 1:5
if percentage(i) ≥= 90
grades{i} = 'A +';
elseif percentage(i) ≥ =80
grades{i} = 'A';
elseif percentage(i) ≥ =70
grades{i} = 'B';
elseif percentage(i) ≥= 60
grades{i} = 'C';
else
grades{i} = 'D';
end
end
```

#### Python

```
grades = []
for p in percentage:
if p > =90:
grades.append("A +")
elif p > =80:
grades.append("A")
elif p > =70:
grades.append("B")
elif p > =60:
grades.append("C")
else:
grades.append("D")
```

### Some extensions

1. Add one invalid mark like 105 and show an error message.
2. Convert marks from integer to float before computing percentage.
3. Store student records using:
   - MATLAB structure array
   - Python dictionary/list of dictionaries
4. Sort students by percentage.
5. Count how many students got each grade.
6. Display pass/fail status using logical data type.
7. Export the report to a CSV file.
8. Add weighted percentage:

```
Math = 40%,   Physics = 30%,   Computer = 30%
```

### Final output format

```
Roll Name Total Percentage Grade
101 Asha 256 85.33 A
102 Ravi 222 74.00 B
103 Neha 275 91.67 A+
104 Karan 146 48.67 D
105 Meera 243 81.00 A
Topper: Neha with 91.67%
Subject Averages:
Math = 74.60
Physics = 76.60
Computer = 77.20
```

CHAPTER 5

# Working with arrays

In the previous chapters, the fundamental concepts of arrays, including their creation, indexing, and basic operations in both Python and MATLAB have been introduced. While these concepts form the foundation of numerical computing, real-world applications often require more advanced techniques for efficient data manipulation.

This chapter focuses on the advanced array operations including transformation, advanced indexing, sorting, combining and splitting arrays, and broadcasting. These operations are essential in scientific computing, data analysis, machine learning, and engineering applications.

Both Python (NumPy) and MATLAB provide powerful tools for handling arrays efficiently. In this chapter, each concept is presented with implementations in both environments, highlighting their similarities and differences to help users transition smoothly between them.

## Array shape and dimension handling

Understanding the shape and dimensions of an array is essential for performing advanced operations. These properties determine how data is organized and how arrays interact during mathematical computations. Both Python and MATLAB provide built-in ways to examine the structure of arrays, which is particularly useful when reshaping, combining, or applying operations across dimensions. Table 5.1 presents a comparison of the commands required to get the information about the rows and columns, dimension, and size of a matrix.

The shape (or size) of an array indicates the number of rows and columns. In Python, shape returns a tuple, whereas MATLAB returns the dimensions directly. The ndim (Python) and ndims (MATLAB) functions indicate the number of dimensions, and size/numel give the total number of elements.

## Array transformation operations

In many computational problems, it is necessary to transform arrays into different forms without altering their data. Array transformation operations allow restructuring of data to suit specific computational requirements such as matrix operations, data analysis, and machine learning. The common transformation opera-

 | HTTPS://DOI.ORG/10.1515/9783112207758-005

TABLE 5.1 Discussion of commands in MATLAB and Python related to array handling.

| | Command | Output |
|---|---|---|
| **Python (NumPy)** | import numpy as np<br>A = np.array([[1, 2, 3],<br>[4, 5, 6]])<br>print(A.shape)<br>print(A.ndim)<br>print(A.size) | (2, 3)<br>2<br>6 |
| **MATLAB** | A = [1 2 3; 4 5 6];<br>size(A)<br>ndims(A)<br>numel(A) | 2 3<br>2<br>6 |

tions include transpose, flattening, and resizing of arrays. Following are the discussion provided for the commands in both the Python and MATLAB.

### Transpose of an array

The transpose of an array interchanges its rows and columns. This operation is widely used in matrix computations and linear algebra. See Table 5.2 for detailed examples.

TABLE 5.2 Comparison of the transpose command along with the output.

| | Command | Output |
|---|---|---|
| **Python (NumPy)** | import numpy as np<br>A = np.array([[1, 2, 3],<br>[4, 5, 6]])<br>print(A.T) | [[1 4]<br>[2 5]<br>[3 6]] |
| **MATLAB** | A = [1 2 3; 4 5 6];<br>A' | $\begin{bmatrix} 1 & 4 \\ 2 & 5 \\ 3 & 6 \end{bmatrix}$ |

Both Python and MATLAB provide simple syntax for transpose. Python uses the attribute .T, whereas MATLAB uses the apostrophe (') operator.

### Flattening an array

Flattening converts a multidimensional array into a one-dimensional array (see examples in Table 5.3). This is useful when data needs to be processed in a linear form.

TABLE 5.3 Presentation of the commands along with the output to flatten an array.

| | Command | Output |
|---|---|---|
| **Python (NumPy)** | A.flatten() | [1 2 3 4 5 6] |
| **MATLAB** | A(:) | 1<br>4<br>2<br>5<br>3<br>6 |

In Python, flatten() returns a new one-dimensional array, whereas in MATLAB, A(:) converts the matrix into a column vector taking the elements from each column one by one.

### Resizing an array

Resizing changes the shape of an array and may alter its structure by repeating or truncating elements if required. See example in Table 5.4.

TABLE 5.4 Presentation of the commands along with the output to resize an array.

| | Command | Output |
|---|---|---|
| **Python (NumPy)** | np.resize(A, (3, 4)) | [[1 2 3 4]<br>[5 6 1 2]<br>[3 4 5 6]] |
| **MATLAB** | reshape(A, 3, 2) | 1 5<br>4 3<br>2 6 |

The resize() command of Python can repeat elements to fit the new shape, whereas reshape() of MATLAB requires the number of elements to remain the same. In MATLAB, the reshape() function requires that the total number of elements remains unchanged.

If the new shape demands more or fewer elements, MATLAB produces an error. In contrast, NumPy provides the resize() function allows reshaping even when the number of elements differs by repeating the original data. This fundamental difference reflects strict enforcement of data consistency in MATLAB, whereas NumPy offers flexibility at the cost of potentially duplicating data.

## Advanced indexing in arrays

In addition to basic indexing and slicing, advanced indexing techniques allow selective access and manipulation of array elements based on conditions or specific index patterns. These techniques are widely used in data filtering, conditional operations, and numerical analysis. Advanced indexing includes boolean indexing and fancy indexing.

### Boolean indexing

Boolean indexing allows selecting elements of an array that satisfy a given condition. This is particularly useful for filtering data. See example in Table 5.5.

TABLE 5.5 Presentation of the commands along with the output to select elements of an array.

| | Command | Output |
|---|---|---|
| **Python (NumPy)** | import numpy as np<br>A = np.array([10, 20, 30, 40, 50])<br>print(A[A > 25]) | [30 40 50] |
| **MATLAB** | A = [10 20 30 40 50];<br>A(A > 25) | 30 40 50 |

Both Python and MATLAB support boolean indexing in a similar way. A logical condition is applied, and elements satisfying the condition are returned.

### Fancy indexing (index-based selection)

Fancy indexing allows accessing elements using specific index positions. This is useful when elements need to be selected in a custom order or pattern. See example with output in Table 5.6.

TABLE 5.6 Presentation of the commands along with the output to extract elements.

| | Command | Output |
|---|---|---|
| **Python (NumPy)** | A = np.array([10, 20, 30, 40, 50])<br>indices = [0, 2, 4]<br>print(A[indices]) | [10 30 50] |
| **MATLAB** | A = [10 20 30 40 50];<br>A([1 3 5]) | 10 30 50 |

Both Python and MATLAB allow indexing using a list of positions. However, Python uses zero-based indexing, whereas MATLAB uses one-based indexing.

### Conditional modification

Advanced indexing can also be used to modify selected elements based on conditions. Boolean indexing can be used not only to select elements but also to modify elements based on a condition. This is a powerful feature that allows updating multiple values in an array without using loops. An example is demonstrated in Table 5.7 for clarity.

TABLE 5.7 Presentation of the commands along with the output to modify elements in an array.

| | Command | Output |
|---|---|---|
| **Python (NumPy)** | A[A > 25] = 0<br>print(A) | [10 20 0 0 0] |
| **MATLAB** | A(A > 25) = 0 | 10 20 0 0 0 |

### Sorting an array

Sorting arranges the elements of an array in ascending or descending order. An example is demonstrated in Table 5.8 to learn the sorting of elements in by defaults ascending order.

TABLE 5.8 Presentation of the commands along with the output to sort elements in an array.

| | Command | Output |
|---|---|---|
| **Python (NumPy)** | import numpy as np<br>A = np.array([40, 10, 30, 20])<br>print(np.sort(A)) | [10 20 30 40] |
| **MATLAB** | A = [40 10 30 20];<br>sort(A) | 10 20 30 40 |

### Sorting in descending order

In order to sort the array in descending order, ans addition keyword needs to be added as shown in Table 5.9 for clarity.

TABLE 5.9 Presentation of the commands along with the output to sort elements in descending order.

| | Command | Output |
|---|---|---|
| **Python (NumPy)** | print(np.sort(A)[::-1]) | [40 30 20 10] |
| **MATLAB** | sort(A, 'descend') | 40 30 20 10 |

Python sorts arrays in ascending order by default, and descending order can be achieved using slicing. MATLAB provides a direct option to sort in descending order using the 'descend' flag.

### Finding indices (searching)

Searching operations help locate the positions of elements that satisfy certain conditions. Table 5.10 provides the details of an example working for the searching of an element with a mentioned condition.

TABLE 5.10 Presentation of the commands along with the output to locate the position of elements.

| | Command | Output |
|---|---|---|
| **Python (NumPy)** | A = np.array([10, 20, 30, 40])<br>print(np.where(A > 20)) | (array([2, 3]),) |
| **MATLAB** | A = [10 20 30 40];<br>find(A > 20) | 3 4 |

Python uses np.where() to return indices of elements satisfying a condition, while MATLAB uses the find() function for the same purpose. Sorting and searching operations are essential for organizing data and are frequently used in algorithms, data analysis, and optimization problems.

## Combining and splitting arrays

In many applications, it is necessary to combine multiple arrays into a single structure or split a large array into smaller parts. These operations are useful in data preprocessing, matrix construction, and handling large datasets. Both Python (NumPy) and MATLAB provide efficient methods for combining and splitting arrays.

### Combining arrays (concatenation)

Concatenation is the process of joining two or more arrays along a specified dimension. See Table 5.11 for concatenating arrays and adding elements in a row as well as a column.

Python uses the concatenate() function with an axis parameter, whereas MATLAB uses simple bracket notation for combining arrays.

TABLE 5.11 Presentation of the commands along with the output to concatenate array elements.

| | Command | Output |
|---|---|---|
| **Python (NumPy)** | import numpy as np<br>A = np.array([[1, 2], [3, 4]])<br>B = np.array([[5, 6], [7, 8]])<br># Vertical stacking<br>C=np.concatenate((A, B), axis=0)<br>print(C) | [[1 2]<br>[3 4]<br>[5 6]<br>[7 8]] |
| **Python (NumPy)** | # Horizontal stacking<br>D=np.concatenate((A, B), axis=1)<br>print(D) | [[1 2 5 6]<br>[3 4 7 8]] |
| **MATLAB** | A = [1 2; 3 4];<br>B = [5 6; 7 8];<br>% Vertical concatenation<br>C=[A; B] | C=<br>1 2<br>3 4<br>5 6<br>7 8 |
| **MATLAB** | % Horizontal concatenation<br>D=[A B] | D=<br>1 2 5 6<br>3 4 7 8 |

## Stacking arrays

Stacking is a convenient way to combine arrays along a new axis. Table 5.12 discusses the commands with examples of stacking arrays in horizontal and vertical directions.

TABLE 5.12 Presentation of the commands along with the output to stack arrays.

| | Command | Output |
|---|---|---|
| **Python (NumPy)** | np.vstack((A,B)) | [[1 2]<br>[3 4]<br>[5 6]<br>[7 8]] |
| | np.hstack((A, B)) | [[1 2 5 6]<br>[3 4 7 8]] |

TABLE 5.12 Presentation of the commands along with the output to stack arrays (*cont.*)

| | Command | Output |
|---|---|---|
| **MATLAB** | [A; B] % vertical<br>[A B] % horizontal | ans =<br>1 2<br>3 4<br>5 6<br>7 8<br>ans =<br>1 2 5 6<br>3 4 7 8 |

Python provides dedicated functions such as vstack() and hstack(), while MATLAB uses the same concatenation syntax for stacking operations.

### Splitting arrays

Splitting divides an array into multiple sub-arrays. Examples of splitting an array are presented in Table 5.13 to demonstrate the working in Python as well as in MATLAB.

TABLE 5.13 Presentation of the commands along with the output to split an array into sub-arrays.

| | Command | Output |
|---|---|---|
| **Python (NumPy)** | A = np.array([[1, 2],<br>[3, 4],<br>[5, 6],<br>[7, 8]])<br># Split into 2 parts vertically<br>print(np.vsplit(A, 2)) | [array([[1, 2],<br>[3, 4]]),<br>array([[5, 6],<br>[7, 8]])] |
| | # Split into 2 parts horizontally<br>print(np.hsplit(A, 2)) | [array([[1],<br>[3],<br>[5],<br>[7]]),<br>array([[2],<br>[4],<br>[6],<br>[8]])] |

TABLE 5.13 Presentation of the commands along with the output to split an array (*cont.*)

| | Command | Output |
|---|---|---|
| **MATLAB** | A = [1 2; 3 4; 5 6; 7 8] | A=<br>1 2<br>3 4<br>5 6<br>7 8 |
| | % Split rows manually<br>A(1:2, :)<br>A(3:4, :) | ans =<br>1 2<br>3 4<br>ans =<br>5 6<br>7 8 |
| | % Split columns manually<br>A(:, 1)<br>A(:, 2) | ans =<br>1<br>3<br>5<br>7<br>ans =<br>2<br>4<br>6<br>8 |

Python provides built-in functions such as vsplit() and hsplit(), whereas MATLAB typically performs splitting using indexing.

The function np.vsplit(A, 2) splits the array into two equal parts along rows (vertical direction). Since the array has four rows, it is divided into two sub-arrays, each containing two rows.

The function np.hsplit(A, 2) splits the array into two equal parts along columns (horizontal direction). Since the array has two columns, it is divided into two sub-arrays, each containing one column.

## Broadcasting in arrays

Broadcasting is a powerful technique that allows arithmetic operations to be performed on arrays of different shapes. Instead of requiring arrays to have the same dimensions, broadcasting automatically expands smaller arrays to match the shape of larger arrays.

This eliminates the need for explicit loops and enables efficient vectorized computations. See examples in Table 5.14.

TABLE 5.14 Presentation of the commands along with the output to understand the concept of broadcasting in arrays.

| | Command | Output |
|---|---|---|
| **Python (NumPy)** | import numpy as np<br>A = np.array([[1, 2, 3],<br>[4, 5, 6]])<br>B = np.array([10, 20, 30])<br>print(A + B) | [[11 22 33]<br>[14 25 36]] |
| **MATLAB** | A = [1 2 3; 4 5 6]<br>B = [10 20 30]<br>A + B | 11 22 33<br>14 25 36 |

Broadcasting (Python) and implicit expansion (MATLAB) allow operations between arrays of different shapes by automatically expanding the smaller array along compatible dimensions. In Python, broadcasting allows arithmetic operations between arrays of different shapes by automatically expanding the smaller array. In MATLAB, this behavior is supported through implicit expansion (introduced in R2016b). In earlier versions of MATLAB, such operations result in an error due to strict dimension matching requirements.

When working with MATLAB, always be aware of the version being used, as broadcasting behavior may differ between older and newer versions.

## Chapter summary

In this chapter, the advanced array operations in Python and MATLAB are discussed including array transformations, advanced indexing, sorting, combining and splitting arrays, and broadcasting. These operations are essential for efficient data manipulation and form the foundation for advanced applications in scientific computing, data analysis, and machine learning. Understanding how to work with array structures and perform operations efficiently allows programmers to write optimized and scalable code in both Python and MATLAB environments.

## Exercise

### Section A: array properties

1. For a given array, determine:
   (a) shape
   (b) number of dimensions
   (c) total number of elements
2. Write equivalent commands in Python and MATLAB to find the size of a matrix.

### Section B: array transformation

3. Find the transpose of a given matrix in:
   (a) Python
   (b) MATLAB
4. Convert a matrix into a one-dimensional array in both environments.

### Section C: advanced indexing

5. Using boolean indexing, extract elements greater than 50 from an array.
6. Select elements at specific positions using:
   (a) Python indexing
   (b) MATLAB indexing
7. Modify all elements less than 10 to zero.

### Section D: sorting and searching

8. Sort a given array in:
   (a) ascending order
   (b) descending order
9. Find the indices of elements greater than a given value.

### Section E: combining and splitting

10. Combine two matrices:
    (a) vertically
    (b) horizontally
11. Split a matrix into two equal parts:
    (a) row-wise
    (b) column-wise

### Section F: broadcasting

12. Perform addition of arrays with different shapes using broadcasting.
13. Explain the rules of broadcasting with an example.

## Computational practice questions with short answers

**Q1. Find the transpose of a 3 × 3 matrix.**
**MATLAB**

```
A = [1 2 3; 4 5 6; 7 8 9];
T = A';
disp(T)
```

**Answer:**
**1 4 7**
**2 5 8**
**3 6 9**

**Python**

```
import numpy as np
A = np.array([[1,2,3],[4,5,6],[7,8,9]])
print(A.T)
```

**Answer:**
**[[1 4 7]**
**[2 5 8]**
**[3 6 9]]**

**Q2. Compute the determinant of a matrix.**
**MATLAB**

```
A = [2 3; 1 4];
det(A)
```

**Answer:**
5

**Python**

```
np.linalg.det(np.array([[2,3],[1,4]]))
```

**Answer:**
5.000000000000001

## Q3. Apply the command to find the inverse of a matrix.
## MATLAB

| | |
|---|---|
| A = [1 2; 3 4]; | **Answer:** |
| inv(A) | −2.0000 1.0000 |
| | 1.5000 −0.5000 |

## Python

| | |
|---|---|
| np.linalg.inv(np.array([[1,2],[3,4]])) | **Answer:** |
| | **array([[−2., 1. ],** |
| | **[ 1.5, −0.5]])** |

## Q4. Solve a set of linear equations.
## MATLAB

| | |
|---|---|
| A = [2 1; 1 3]; | **Answer:** |
| b = [5; 6]; | x = |
| x = A\b | 1.8000 |
| | 1.4000 |

## Python

| | |
|---|---|
| A = np.array([[2,1],[1,3]]) | **Answer:** |
| b = np.array([5,6]) | [1.8 1.4] |
| x = np.linalg.solve(A,b) | |
| print(x) | |

## Q5. Calculate the eigenvalues of a square matrix.
## MATLAB

| | |
|---|---|
| eig([2 0; 0 3]) | **Answer:** |
| | ans = |
| | 2 |
| | 3 |

**Python**

| | |
|---|---|
| np.linalg.eigvals(np.array([[2,0],[0,3]])) | **Answer:**<br>array([2., 3.]) |

**Q6. Verify the array slicing using the provided command.**
**MATLAB**

| | |
|---|---|
| A = 1:10;<br>A(3:2:9) | **Answer:**<br>ans =<br>3 5 7 9 |

**Python**

| | |
|---|---|
| A = np.arange(1,11)<br>print(A[2:9:2]) | **Answer:**<br>[3 5 7 9] |

**Q7. Investigate the command to obtain the reverse of an array.**
**MATLAB**

| | |
|---|---|
| A = [1 2 3 4];<br>fliplr(A) | **Answer:**<br>4 3 2 1 |

**Python**

| | |
|---|---|
| A = np.array([1,2,3,4])<br>print(A[::-1]) | **Answer:**<br>[4 3 2 1] |

**Q8. For a given matrix A, verify the extraction of the diagonal elements.**
**MATLAB**

| | |
|---|---|
| A = [1 2; 3 4];<br>diag(A) | **Answer:**<br>1<br>4 |

**Python**

| | |
|---|---|
| np.diag(np.array([[1,2],[3,4]])) | **Answer:**<br>Array([1, 4]) |

Q9. **Obtain the results of element-wise multiplication for the two defined matrices.**

**MATLAB**

```
A = [1 2 3];
B = [4 5 6];
A.*B
```

**Answer:**
4 10 18

**Python**

```
A = np.array([1,2,3])
B = np.array([4,5,6])
print(A*B)
```

**Answer:**
[4 10 18]

Q10. **Perform the matrix multiplication on the two matrices.**

**MATLAB**

```
A = [1 2; 3 4];
B = [5 6; 7 8];
A*B
```

**Answer:**
19 22
43 50

**Python**

```
A = np.array([[1,2],[3,4]])
B = np.array([[5,6],[7,8]])
print(A @ B)
```

**Answer:**
[[19 22]
[43 50]]

### Q11. Perform the broadcasting addition.

MATLAB

| Code | Answer: |
|---|---|
| A = [1;2;3]; | 11 21 31 |
| B = [10 20 30]; | 12 22 32 |
| bsxfun(@plus, A, B) | 13 23 33 |

Python

| Code | Answer: |
|---|---|
| A = np.array([[1],[2],[3]]) | [[11 21 31] |
| B = np.array([10,20,30]) | [12 22 32] |
| print(A + B) | [13 23 33]] |

### Q12. Reshape the defined array.

MATLAB

| Code | Answer: |
|---|---|
| reshape(1:6,2,3) | 1 3 5 |
| | 2 4 6 |

Python

| Code | Answer: |
|---|---|
| np.arange(1,7).reshape(2,3) | array([[1, 2, 3], |
| | [4, 5, 6]]) |

### Q13. Investigate the concept of the flatten array.

MATLAB

| Code | Answer: |
|---|---|
| A = [1 2; 3 4]; | 1 |
| A(:) | 2 |
| | 3 |
| | 4 |

### Python

| Code | Answer: |
|---|---|
| import numpy as np<br>A = np.array([[1, 2], [3, 4]])<br>print(A.flatten()) | [1 2 3 4] |

### Q14. Verify the results of the logical indexing.

### MATLAB

| Code | Answer: |
|---|---|
| A = [2 6 8 3];<br>A(A>5) | 6 8 |

### Python

| Code | Answer: |
|---|---|
| import numpy as np<br>A = np.array([2, 6, 8, 3])<br>result = A[A > 5]<br>print(result) | [6,8] |

### Q15. Perform the required operation to replace the negative values by zero in the defined matrix.

### MATLAB

| Code | Answer: |
|---|---|
| A = [−1 2 −3 4];<br>A(A<0)=0 | 0 2 0 4 |

### Python

| Code | Answer: |
|---|---|
| A = np.array([−1,2,−3,4])<br>A[A<0] = 0<br>print(A) | [ 0 2 0 4] |

Q16. **Calculate the sum of elements of a given list.**
**MATLAB**

| | |
|---|---|
| sum([1 2 3 4]) | **Answer:**<br>10 |

**Python**

| | |
|---|---|
| np.sum([1,2,3,4]) | **Answer:**<br>10 |

Q17. **Calculate the row-wise sum of the given matrix.**
**MATLAB**

| | |
|---|---|
| sum([1 2; 3 4],2) | **Answer:**<br>3<br>7 |

**Python**

| | |
|---|---|
| np.sum([[1,2],[3,4]], axis=1) | **Answer:**<br>array([3, 7]) |

Q18. **Calculate the column-wise sum of the given matrix.**
**MATLAB**

| | |
|---|---|
| mean([1 2; 3 4]) | **Answer:**<br>2 3 |

**Python**

| | |
|---|---|
| np.mean([[1,2],[3,4]], axis=0) | **Answer:**<br>array([2., 3.]) |

Q19. **Calculate the dot product of the given matrix.**
**MATLAB**

| | |
|---|---|
| dot([1 2 3],[4 5 6]) | **Answer:**<br>32 |

**Python**

| | |
|---|---|
| np.dot([1,2,3],[4,5,6]) | **Answer:**<br>32 |

Q20. **Generate the cumulative sum for a defined list of numbers.**
**MATLAB**

| | |
|---|---|
| cumsum([1 2 3 4]) | **Answer:**<br>1 3 6 10 |

**Python**

| | |
|---|---|
| np.cumsum([1,2,3,4]) | **Answer:**<br>array([ 1, 3, 6, 10]) |

Q21. **Find the maximum element of a list.**
**MATLAB**

| | |
|---|---|
| max([3 7 2 9]) | **Answer:**<br>9 |

**Python**

| | |
|---|---|
| np.max([3,7,2,9]) | **Answer:**<br>9 |

Q22. **Sort the element of an array.**
**MATLAB**

| | |
|---|---|
| sort([4 1 3 2]) | **Answer:**<br>1 2 3 4 |

**Python**

| | |
|---|---|
| np.sort([4,1,3,2]) | **Answer:**<br>array([1, 2, 3, 4]) |

Q23. **Extract the unique elements from a vector.**
**MATLAB**

| | |
|---|---|
| unique([1 2 2 3 3 3]) | **Answer:**<br>1 2 3 |

**Python**

| | |
|---|---|
| np.unique([1,2,2,3,3,3]) | **Answer:**<br>array([1, 2, 3]) |

Q24. **Calculate the rank of a matrix.**
**MATLAB**

| | |
|---|---|
| rank([1 2; 2 4]) | **Answer:**<br>1 |

**Python**

| | |
|---|---|
| np.linalg.matrix_rank([[1,2],[2,4]]) | **Answer:**<br>1 |

Q25. **Obtain the trace of a matrix.**
**MATLAB**

| | |
|---|---|
| trace([1 2; 3 4]) | **Answer:** 5 |

**Python**

| | |
|---|---|
| np.trace([[1,2],[3,4]]) | **Answer:** 5 |

## Mini project: multivariate data analysis using advanced array operations

### Objective

Analyze a multivariate dataset (matrix form) using:
- Matrix algebra
- Broadcasting
- Statistical operations
- Logical indexing
- Transformations

You will implement this in MATLAB and Python (NumPy).

**Problem statement**
You are given a dataset of students' scores in four subjects:

$$D = \begin{bmatrix} 85 & 78 & 92 & 88 \\ 76 & 81 & 85 & 80 \\ 90 & 95 & 93 & 89 \\ 60 & 65 & 70 & 72 \\ 88 & 84 & 91 & 87 \end{bmatrix}$$

**Tasks:**

### Task 1: Basic matrix operations

- Compute:
  - Mean of each subject (column-wise)
  - Total score per student (row-wise)

Concepts: Mean, summation

Hint:
- MATLAB: mean(D), sum(D,2)
- Python: np.mean(D, axis=0), np.sum(D, axis=1)

### Task 2: Normalization (broadcasting)

Normalize scores using:

$$Z = \frac{D - \mu}{\sigma}$$

Concepts: Mean, standard deviation, broadcasting

Hint:
- Compute mean & std first
- Subtract vector from matrix (broadcasting)

### Task 3: Top performers (logical indexing)

Find students with average score >85.
Concepts: Logical masking

Hint:
- Compute row mean
- Apply condition: >85

### Task 4: Rank students

Rank students based on total marks.
Concepts: Sorting, indexing

Hint:
- MATLAB: sort( ..., 'descend')
- Python: np.argsort(-total)

### Task 5: Covariance matrix

Compute covariance matrix of subjects.
Concept:

$$\mathrm{Cov}(X, Y) = \frac{1}{n-1}(X-\mu)^T(X-\mu)$$

Hint:
- MATLAB: cov(D)
- Python: np.cov(D, rowvar=False)

### Task 6: Principal component analysis (PCA – basic)

- Compute eigenvalues and eigenvectors of covariance matrix.

Concepts: Eigendecomposition

Hint:
- MATLAB: eig(C)
- Python: np.linalg.eig(C)

### Task 7: Reshape and flatten

Convert dataset into:
- Single vector
- Back to matrix

Hint:
- MATLAB: D(:)
- Python: D.flatten()

### Task 8: Data filtering

Replace all scores below 70 with NaN or 0.
Concepts: Conditional replacement

Hint:
- MATLAB: D(D<70)=0
- Python: D[D<70]=0

### Task 9: Weighted scores

Apply weight vector:

$$W = [0.2, 0.3, 0.3, 0.2]$$

Compute final score:

$$S = D \cdot W^T$$

Concepts: Matrix multiplication

Hint:
- MATLAB: D*W'
- Python: D @ W

### Task 10: Correlation matrix

Compute correlation between subjects.

Hint:
- MATLAB: corrcoef(D)
- Python: np.corrcoef(D, rowvar=False)

### Extension (advanced)

- Visualize data (heatmap)
- Compare normalized versus original data
- Identify most correlated subjects
- Implement PCA projection manually

CHAPTER 6

# Conditional statements and loops

In the previous chapters, the concept of data structures, array operations, and numerical computations in both Python and MATLAB were discussed. However, real-world problems often require programs to make decisions and execute tasks repeatedly based on specific conditions.

This chapter introduces control flow, which determines the order in which statements are executed in a program. It includes conditional statements for decision-making and loops for repetitive execution.

**Conditional statements** are the tool that enables a program to evaluate conditions and execute different blocks of code. **Loops** enable repeated execution of a set of instructions until a condition is satisfied.

Understanding these concepts is essential for developing efficient algorithms and hence solving computational problems in scientific computing, data analysis, and engineering applications. Both Python and MATLAB provide structured ways to implement the control flow with some differences in syntax and execution style that is explored with details in this chapter.

## Conditional statements (decision-making)

Conditional statements are used to execute specific blocks of code based on whether a condition is true or false. These statements form the basis of decision-making in programming. The simplest form of the decision-making statement is using if condition. It allows to check whether the condition is true or false. The statement followed the if condition is executed if the condition is true. An example is compared for both Python and MATLAB to check whether the variable has a value greater than 5 in Table 6.1. The string will be printed when the condition is true.

The if statement checks whether the condition is true. If the condition is satisfied, the corresponding block of code is executed. Otherwise, the block is skipped.

Python uses indentation to define blocks of code, whereas MATLAB uses the keyword end to terminate conditional blocks. A common mistake in Python is incorrect indentation, which leads to syntax errors. In MATLAB, forgetting the end keyword results in incomplete conditional statements.

 | HTTPS://DOI.ORG/10.1515/9783112207758-006

TABLE 6.1 Comparison of commands with examples to use the conditional statement on MATLAB and Python.

| | Command | Output |
|---|---|---|
| **Python (NumPy)** | x=10<br>if x>5:<br>    print("x is greater than 5") | x is greater than 5 |
| **MATLAB** | x=10;<br>if x>5<br>    disp('x is greater than 5')<br>End | x is greater than 5 |

### If-else and elif statements

In many situations, a program must choose between multiple alternatives based on a condition. The **if-else** statement allows execution of one block of code when a condition is true and another block when the condition is false.

For handling multiple conditions, Python provides the elif statement, which allows checking additional conditions sequentially. The example considered above is further extended in Table 6.2, in which if the condition is not true a message will be displayed mentioning that the value is less than or equal to 5.

TABLE 6.2 Comparison of if-else commands with examples on MATLAB and Python.

| | Command | Output |
|---|---|---|
| **Python (NumPy)** | x=3<br>if x>5:<br>    print("x is greater than 5")<br>else:<br>    print("x is less than or equal to 5") | x is less than or equal to 5 |
| **MATLAB** | x=3;<br>if x>5<br>    disp('x is greater than 5')<br>else<br>    disp('x is less than or equal to 5')<br>end | x is less than or equal to 5 |

The if-else statement provides two execution paths. If the condition is true, the if block is executed; otherwise, the else block is executed. Both Python and MATLAB support if-else loop for decision-making. Python uses indentation to define code blocks, whereas MATLAB uses the end keyword to terminate the conditional structure.

**The elif statement** allows checking multiple conditions in sequence. Once a condition is satisfied, the corresponding block is executed, and the remaining conditions are skipped. See Table 6.3 for detail example.

TABLE 6.3 Examples of using multiple conditional statements in both MATLAB and Python.

| | Command | Output |
|---|---|---|
| **Python (NumPy)** | marks=75<br>if marks>=90:<br>    print("Grade A")<br>elif marks>=75:<br>    print("Grade B")<br>elif marks>=50:<br>    print("Grade C")<br>else:<br>    print("Fail") | Grade B |
| **MATLAB** | marks=75;<br>if marks>=90<br>    disp('Grade A')<br>elseif marks>=75<br>    disp('Grade B')<br>elseif marks>=50<br>    disp('Grade C')<br>else<br>    disp('Fail')<br>End | Grade B |

The elif (Python) or elseif (MATLAB) statement allows checking multiple conditions sequentially. Only the first condition that evaluates to true is executed.

## Nested conditions and logical operators

In many practical problems, a single condition is not sufficient to make a decision. Nested conditions and logical operators allow combining multiple conditions and creating more complex decision structures. Nested conditions involve placing one conditional statement inside another, while logical operators are used to combine multiple conditions into a single expression.

### Nested conditions

A nested condition is an if statement placed inside another if or else block. This allows checking additional conditions after an initial condition is satisfied. Table 6.4 demonstrates the example with nested condition.

TABLE 6.4 Examples of using nested if condition in both MATLAB and Python.

| | Command | Output |
|---|---|---|
| **Python (NumPy)** | x=10<br>if x>5:<br>    if x<15:<br>        print("x is between 5 and 15") | x is between 5 and 15 |
| **MATLAB** | x=10;<br>if x>5<br>    if x<15<br>        disp('x is between 5 and 15')<br>    end<br>end | x is between 5 and 15 |

Both Python and MATLAB support nested conditions. Python relies on indentation to define nesting, whereas MATLAB uses the end keyword for each conditional block.

### Logical operators

Logical operators are used to combine multiple conditions into a single statement. These operators are useful when decisions depend on more than one condition.

### Common logical operators

The common logical operators in every programming language include the and, or, and not operators. These operators help in evaluating multiple conditions simultaneously by determining whether one or more conditions are true or false. Combining logical operators with conditional statements allows writing compact and efficient decision-making logic without deeply nested structures. The symbols used for the logical operators are presented in Table 6.5.

TABLE 6.5 Logical operator symbols used in MATLAB and Python.

| Operation | Python | MATLAB |
|---|---|---|
| **AND** | and | && |
| **OR** | or | \|\| |
| **NOT** | not | ~ |

Logical operators are used to combine multiple conditions. The AND operator requires both conditions to be true, OR requires at least one condition to be true, and NOT reverses the condition.

Table 6.6 provides the example compared for the context of both Python and MATLAB giving insight of evaluating more than one condition and giving the decision accordingly.

TABLE 6.6 Examples of evaluating more than one condition in both MATLAB and Python.

| | Command | Output |
|---|---|---|
| **Python (NumPy):** Example Using AND | x=10<br>if x>5 and x<15:<br>    print("x is between 5 and 15") | x is between 5 and 15 |
| **Python (NumPy):** Example Using OR | x=20<br>if x<10 or x>15:<br>    print("Condition satisfied") | Condition satisfied |

TABLE 6.6 Examples of evaluating more than one condition in both MATLAB (*cont.*)

<table>
<tr><th></th><th>Command</th><th>Output</th></tr>
<tr><td>Python (NumPy): Example Using NOT</td><td>x=5<br>if not(x>10):<br>    print("x is not greater than 10")</td><td>x is not greater than 10</td></tr>
<tr><td>MATLAB: Example Using AND</td><td>x=10;<br>if x>5 && x<15<br>    disp('x is between 5 and 15')<br>end</td><td>x is between 5 and 15</td></tr>
<tr><td>MATLAB: Example Using OR</td><td>x=20;<br>if x<10 || x>15<br>    disp('Condition satisfied')<br>end</td><td>Condition satisfied</td></tr>
<tr><td>MATLAB: Example Using NOT</td><td>x=5;<br>if~(x>10)<br>    disp('x is not greater than 10')<br>end</td><td>x is not greater than 10</td></tr>
</table>

Python uses keywords such as and, or, and not for logical operations, whereas MATLAB uses symbols like &&, ||, and ~. The functionality remains similar, but syntax differs. In MATLAB, && and || are used for scalar logical operations, while & and | are used for element-wise logical operations on arrays. Logical operators allow combining multiple conditions into a single expression. This reduces the need for deeply nested conditional statements and improves code readability and efficiency.

## For loop

A **for** loop is used to execute a block of code repeatedly for a fixed number of iterations. It is commonly used when the number of repetitions is known in advance. Both Python and MATLAB support for loops, but their syntax and structure differ significantly.

Table 6.7 compares Python and MATLAB in terms of syntax and the iteration requirements to proceed with loop involving sequence.

TABLE 6.7 Comparison of the loops in both Python and MATLAB for syntax.

| | Command | Output | Explanation |
|---|---|---|---|
| **Python:** The **for** loop iterates over a sequence such as a list, range, or array. | for i in range(5):<br>    print(i) | 0<br>1<br>2<br>3<br>4 | The function range(5) generates a sequence from 0 to 4. The loop variable i takes each value in the sequence, and the print statement is executed in each iteration. |
| MATLAB: The **for** loop iterates over a defined range of values. | for i=0:4<br>    disp(i)<br>end | 0<br>1<br>2<br>3<br>4 | MATLAB uses the colon operator (start:end) to define the loop range, and the loop variable takes each value sequentially. |

Iteration refers to the repeated execution of a set of instructions. for loops are used when the number of iterations is known in advance, whereas while loops are used when execution depends on a condition.

### Loop with step size

The example of creating the loops along with the details of the iterations can be understood from Table 6.8.

TABLE 6.8 Presentation of the commands of Python and MATLAB for iterations.

| | Command | Output | Explanation |
|---|---|---|---|
| **Python** | for i in range (0, 10, 2):<br>    print(i) | 0<br>2<br>4<br>6<br>8 | Step size determines the increment between values in the loop. Python uses range(start, stop, step), while MATLAB uses start:step:end to control iteration. |
| **MATLAB** | for i=0:2:8<br>    disp(i)<br>end | 0<br>2<br>4<br>6<br>8 | |

In Python, range(start, stop, step) defines the loop sequence, while MATLAB uses start:step:end syntax. Both achieve similar iteration patterns.

**Looping through arrays**

The arrays can also be printed using the concepts of loops. The Python directly iterates over elements of a list or array, whereas MATLAB assigns each element of the array sequentially to the loop variable. In Python, loops are often combined with built-in functions and vectorized operations, while MATLAB typically emphasizes vectorized computations over explicit loops for efficiency. See Table 6.9 for understanding with example.

TABLE 6.9 Comparison of the commands to print arrays in both MATLAB and Python.

| | Command | Output |
|---|---|---|
| **Python** | A=[10, 20, 30]<br>for value in A:<br>    print(value) | 10<br>20<br>30 |
| **MATLAB** | A=[10 20 30];<br>for value=A<br>    disp(value)<br>end | 10<br>20<br>30 |

## While loop

A while loop is used to execute a block of code repeatedly as long as a given condition remains true. Unlike the for loop, the number of iterations is not fixed in advance and depends on the condition. **While** loops are useful in situations where the termination condition is not known beforehand.

In Python, the while loop continues execution as long as the specified condition evaluates to true. Both Python and MATLAB use while loops for condition-based iteration. Python uses indentation to define the loop body, whereas MATLAB uses the end keyword to terminate the loop. It is essential to update the loop variable inside the while loop; otherwise, the loop may run indefinitely (infinite loop). An infinite loop occurs when the terminating condition of a loop is never met. This can cause a program to run indefinitely and must be avoided by ensuring proper update of loop variables. The example of while loop is provided in Table 6.10 for clarity.

TABLE 6.10 Comparison of the execution of while loop in Python and MATLAB.

| | Command | Output | Explanation |
|---|---|---|---|
| **Python** | i=0<br>while i<5:<br>    print(i)<br>    i+=1 | 0<br>1<br>2<br>3<br>4 | The loop starts with i=0 and continues as long as i is less than 5. In each iteration, the value of i is printed and then incremented by 1. When i becomes 5, the condition becomes false, and the loop stops. |
| **MATLAB** | i=0;<br>while i<5<br>    disp(i)<br>    i=i+1;<br>end | 0<br>1<br>2<br>3<br>4 | In MATLAB, the while loop also executes repeatedly based on a condition, similar to Python. |

### Infinite loop

While loops are best suited for problems where the number of iterations depends on a condition rather than a predefined range. But some time, it results in the infinite loop as in examples discussed in Table 6.11.

TABLE 6.11 Examples that result in the infinite loop.

| | Command | Output | Explanation |
|---|---|---|---|
| **Python** | i=0<br>while i<5:<br>    print(i) | | This loop will run indefinitely because i is never updated. |
| **MATLAB** | i=0;<br>while i<5<br>    disp(i)<br>end | | |

## Loop control statements

Loop control statements are used to alter the normal flow of loops. They allow a program to exit a loop prematurely, skip certain iterations, or act as a placeholder where no action is required.

The commonly used loop control statements are break, continue, and pass (Python only).

### Break statement

The break statement is used to terminate a loop immediately when a specified condition is met. Both Python and MATLAB use the break statement to exit a loop immediately when a condition is satisfied. See example in Table 6.12.

TABLE 6.12 Demonstration of the usage of break statement inside the loop.

| | Command | Output | Explanation |
|---|---|---|---|
| **Python** | for i in range(5):<br>  if i==3:<br>    break<br>    print(i) | 0<br>1<br>2 | The break statement terminates the loop immediately when the specified condition is met, preventing further iterations. |
| **MATLAB** | for i=0:4<br>  if i==3<br>    break<br>    end<br>    disp(i)<br>end | 0<br>1<br>2 | |

### Continue statement

The continue statement skips the current iteration and moves to the next iteration of the loop. The detailed example is presented in Table 6.13.

Both Python and MATLAB support the continue statement to skip specific iterations without terminating the loop.

TABLE 6.13 Examples demonstrating the continue statement inside the loop.

| | Command | Output | Explanation |
|---|---|---|---|
| **Python** | for i in range(5):<br>  if i==2:<br>    continue<br>    print(i) | 0<br>1<br>3<br>4 | The continue statement skips the current iteration and moves to the next iteration of the loop without executing the remaining statements. |
| **MATLAB** | for i=0:4<br>  if i==2<br>    continue<br>  end<br>  disp(i)<br>end | 0<br>1<br>3<br>4 | |

### Pass statement (Python only)

The pass statement is used as a placeholder in Python where a statement is syntactically required but no action needs to be performed.

**Example:**

```
for i in range(3):
    if i = = 1:
        pass
    print(i)
```

**Output:**

```
0
1
2
```

### Summary comparison

A summary of the loop control statements is provided in Table 6.14 with the discussion on the purpose of the control statement. It compares the purpose of the available or not existing commands in both Python and MATLAB.

Loop control statements provide flexibility in managing loop execution and are particularly useful in handling special conditions within iterative processes.

TABLE 6.14 Comparison of the loop control statements of Python and MATLAB.

| Statement | Python | MATLAB | Purpose |
|---|---|---|---|
| **break** | break | break | Exit loop |
| **continue** | continue | continue | Skip iteration |
| **pass** | pass | Not available | Placeholder |

## Common mistakes to avoid

- Incorrect indentation in Python conditional and loop statements
- Forgetting the end keyword in MATLAB
- Using multiple if statements instead of elif/elseif
- Forgetting to update variables in while loops (leading to infinite loops)
- Misusing logical operators in conditions

## PYTHON versus MATLAB summary

Both Python and MATLAB provide similar control flow constructs, but differ in syntax and implementation. Python emphasizes readability through indentation, whereas MATLAB uses explicit keywords to define structure. Table 6.15 summarizes the key differences between Python and MATLAB in terms of control flow structures.

TABLE 6.15 Summary of the control flow commands for both Python and MATLAB.

| Feature | Python | MATLAB |
|---|---|---|
| **Block structure** | Indentation | end keyword |
| **Multiple conditions** | elif | elseif |
| **Logical operators** | and, or, not | &&, \|\|, ~ |
| **for loop syntax** | range() | start:end |
| **while loop** | condition-based | condition-based |
| **Loop control** | break, continue, pass | break, continue |

## Summary

In this chapter, the fundamental concepts of control flow in programming are discussed. It includes the conditional statements and loops in both Python and MATLAB. Conditional statements such as if, if-else, and multiple conditions using elif/elseif enable decision-making in programs are discussed with examples.

Nested conditions and logical operators allow combining multiple conditions to form complex decision structures. Iterative conditional checking has been discussed using for and while loops, that enable repeated execution of statements, making programs efficient and dynamic. Loop control statements such as break and continue provide flexibility in managing loop execution. These statements are discussed for MATLAB while Python also includes the pass statement as a placeholder. Understanding these control mechanisms is essential for developing algorithms and solving computational problems effectively in both Python and MATLAB environments.

## Working examples

**Example 1: Write code in both Python and MATLAB to find even numbers from 1 to 11.**

**Python:**

```
for i in range(1, 11):
    if i % 2==0:
        print(i)
```

**MATLAB:**

```
for i=1:10
    if mod(i,2)==0
        disp(i)
    end
end
```

**Example 2: Write a program to find the factorial of a number using a loop.**

**Python**

```
n=5
fact=1
for i in range(1, n+1):
    fact *=i
print(fact)
```

**Output:**

```
120
```

**MATLAB**

```
n=5;
fact=1;
for i=1:n
        fact=fact * i;
end
disp(fact)
```

**Output:**

```
120
```

This example demonstrates the use of a for loop to perform repeated multiplication.

**Example 3: Write a program to find the maximum value in a list/array.**

**Python**

```
A=[10, 45, 23, 67, 34]
max_val=A[0]
for value in A:
        if value>max_val:
            max_val = value
print(max_val)
```

**Output:**

```
67
```

**MATLAB**

```
A=[10 45 23 67 34];
max_val=A(1);
for i=A
        if i>max_val
            max_val=i;
        end
end
disp(max_val)
```

**Output:**

```
67
```

## Exercise

### Section A: conditional statements

1. Write a program in Python and MATLAB to:
   - check whether a number is positive, negative, or zero
2. Write a program to:
   - determine whether a number is even or odd
3. Write a program using multiple conditions to assign grades based on marks:
   - marks≥90 → Grade A
   - marks≥75 → Grade B
   - marks≥50 → Grade C
   - otherwise → Fail

### Section B: logical operators

4. Write a program to check whether a number lies between 10 and 50 using logical operators.
5. Write a program to check whether a number is divisible by both 2 and 3.

### Section C: for loop

6. Write a program to print numbers from 1 to 10 using:
   (a) Python
   (b) MATLAB
7. Write a program to calculate the sum of first n natural numbers using a for loop.

### Section D: while loop

8. Write a program using a while loop to:
   - print numbers from 1 to 10
9. Write a program to find the factorial of a number using a while loop.

### Section E: loop control statements

10. Write a program to stop printing numbers when a number greater than 5 is encountered (use break)
11. Write a program to skip even numbers and print only odd numbers (use continue)

## Computational practice questions with short answers

1. **Check whether a number is positive or negative.**

**MATLAB**

```
x=-4;
if x>0
    disp('Positive')
else
    disp('Negative')
end
```

**Python**

```
x=-4
if x>0:
    print("Positive")
else:
    print("Negative")
```

2. **Check whether a number is even or odd.**

**MATLAB**

```
n=7;
if mod(n,2)==0
    disp('Even')
else
    disp('Odd')
end
```

**Python**

```
n=7
if n % 2==0:
    print("Even")
else:
    print("Odd")
```

### 3. Assign grades using multiple conditions.

**MATLAB**

```
marks=85;
if marks>=90
    disp('Grade A')
elseif marks>=75
    disp('Grade B')
else
    disp('Grade C')
end
```

**Python**

```
marks=85
if marks>=90:
    print("Grade A")
elif marks>=75:
    print("Grade B")
else:
    print("Grade C")
```

### 4. Check whether a number lies between 10 and 50.

**MATLAB**

```
x=25;
if x>=10 && x<=50
    disp('Inside Range')
end
```

**Python**

```
x=25
if x>= 10 and x<=50:
    print("Inside Range")
```

### 5. Check divisibility by both 2 and 3.

**MATLAB**

```
n=12;
if mod(n,2)==0 && mod(n,3)==0
    disp('Divisible')
end
```

**Python**

```
n=12
if n % 2==0 and n % 3==0:
    print("Divisible")
```

### 6. Print numbers from 1 to 10.

**MATLAB**

```
for i=1:10
    disp(i)
end
```

**Python**

```
for i in range(1,11):
    print(i)
```

### 7. Print even numbers from 1 to 20.

**MATLAB**

```
for i=1:20
    if mod(i,2)==0
      disp(i)
    end
end
```

### Python

```
for i in range(1,21):
    if i % 2==0:
      print(i)
```

### 8. Find the sum of first 10 natural numbers

### MATLAB

```
sum1=0;
for i=1:10
    sum1=sum1+i;
end
disp(sum1)
```

### Python

```
sum1=0
for i in range(1,11):
    sum1+=i
print(sum1)
```

### 9. Find factorial of a number.

### MATLAB

```
fact=1;
for i=1:5
    fact=fact * i;
end
disp(fact)
```

### Python

```
fact=1
for i in range(1,6):
    fact *=i
print(fact)
```

## 10. Print multiplication table of 5.

**MATLAB**

```
for i=1:10
    disp(5*i)
end
```

**Python**

```
for i in range(1,11):
    print(5*i)
```

## 11. Find maximum element in an array.

**MATLAB**

```
A=[12 45 7 89 23];
max_val = A(1);
for i=A
    if i>max_val
      max_val=i;
    end
end
disp(max_val)
```

**Python**

```
A=[12,45,7,89,23]
max_val=A[0]
for i in A:
    if i>max_val:
      max_val=i
print(max_val)
```

### 12. Find minimum element in an array.
**MATLAB**

```
A=[12 45 7 89 23];
min_val=A(1);
for i=A
    if i<min_val
    min_val=i;
    end
end
disp(min_val)
```

**Python**

```
A=[12,45,7,89,23]
min_val=A[0]
for i in A:
    if i<min_val:
        min_val=i
print(min_val)
```

### 13. Print numbers using while loop.
**MATLAB**

```
i=1;
while i<=5
    disp(i)
    i=i+1;
end
```

**Python**

```
i=1
while i<=5:
  print(i)
    i+=1
```

14. **Find factorial using while loop.**
**MATLAB**

```
n=5;
fact=1;
i=1;
while i<=n
    fact=fact * i;
    i=i+1;
end
disp(fact)
```

**Python**

```
n=5
fact=1
i=1
while i<=n:
    fact *=i
    i+=1
print(fact)
```

15. **Use break statement to stop loop.**
**MATLAB**

```
for i=1:10
    if i==6
        break
    end
    disp(i)
end
```

**Python**

```
for i in range(1,11):
    if i==6:
        break
    print(i)
```

### 16. Use continue statement to skip 5.

**MATLAB**

```
for i=1:10
    if i==5
    continue
    end
    disp(i)
end
```

**Python**

```
for i in range(1,11):
    if i==5:
        continue
    print(i)
```

### 17. Print only odd numbers.

**MATLAB**

```
for i=1:10
    if mod(i,2)==0
    continue
    end
    disp(i)
end
```

**Python**

```
for i in range(1,11):
    if i % 2==0:
        continue
    print(i)
```

### 18. Count positive numbers in an array.

**MATLAB**

```
A=[-2 4 6 -1 8];
count=0;
for i=A
    if i>0
        count=count+1;
    end
end
disp(count)
```

**Python**

```
A=[-2,4,6,-1,8]
count=0
for i in A:
    if i>0:
        count+=1
print(count)
```

### 19. Find sum of even numbers from 1 to 20

**MATLAB**

```
sum1=0;
for i=1:20
    if mod(i,2)==0
        sum1=sum1+i;
    end
end
disp(sum1)
```

**Python**

```
sum1=0
for i in range(1,21):
    if i % 2==0:
        sum1 + = i
print(sum1)
```

20. **Print square of numbers from 1 to 5.**
**MATLAB**

```
for i=1:5
    disp(i^2)
end
```

**Python**

```
for i in range(1,6):
    print(i**2)
```

21. **Find cube of numbers from 1 to 5.**
**MATLAB**

```
for i=1:5
    disp(i^3)
end
```

**Python**

```
for i in range(1,6):
    print(i**3)
```

22. **Count vowels in a string.**
**MATLAB**

```
str='education';
count=0;
for ch=str
    if any(ch=='aeiou')
      count=count+1;
    end
end
disp(count)
```

### Python

```
str1="education"
count=0
for ch in str1:
    if ch in "aeiou":
        count+=1
print(count)
```

### 23. Print Fibonacci series up to 10 terms.

### MATLAB

```
a=0;
b=1;
disp(a)
disp(b)
for i=1:8
    c=a+b;
    disp(c)
    a=b;
    b=c;
end
```

### Python

```
a=0
b=1
print(a)
print(b)
for i in range(8):
    c=a+b
    print(c)
    a=b
    b=c
```

## 24. Check whether a number is an Armstrong number.

**MATLAB**

```
n=153;
temp=n;
sum1=0;
while temp>0
    r=mod(temp,10);
    sum1=sum1+r^3;
    temp=floor(temp/10);
end
if sum1==n
    disp('Armstrong Number')
else
    disp('Not Armstrong')
end
```

**Python**

```
n=153
temp=n
sum1=0
while temp>0:
    r=temp % 10
    sum1+=r**3
    temp = temp // 10
if sum1==n:
    print("Armstrong Number")
else:
    print("Not Armstrong")
```

25. **Find the greatest common divisor (GCD) of two numbers.**
**MATLAB**

```
a=24;
b=36;
while b~=0
    r=mod(a,b);
    a=b;
    b=r;
end
disp(a)
```

**Python**

```
a=24
b=36
while b !=0:
    r=a % b
    a=b
    b=r
print(a)
```

## Mini project: student result management system using conditional statements and loops

This mini project helps students apply concepts of:
- if, if-else, elseif/elif
- logical operators
- for loop
- while loop
- loop control statements

### Objective

Develop a simple **student result management system** that:
1. Accepts marks of students
2. Calculates total and average marks
3. Assigns grades

4. Counts passed and failed students
5. Displays topper marks

### Problem statement

Write a program in MATLAB and Python to:
- Input marks of five students
- Calculate:
  - Total marks
  - Average marks
  - Highest marks
- Assign grades:
  - ≥90 → Grade A
  - ≥75 → Grade B
  - ≥50 → Grade C
  - < 50 → Fail
- Count total passed students

CHAPTER 7

# Functions in MATLAB and Python

While writing out a program for an intended task many times a particular set of statements needs to be executed repeatedly. Writing the same code again and again increases the length of the program and makes it difficult to manage. To overcome this problem, programmers use the concept of functions.

*A function is a self-contained block of code designed to perform a specific task.*

Functions help in dividing a large program into smaller and manageable parts by writing the repeated statements collectively with a meaningful name. This modular approach improves readability, debugging, maintenance, and reusability of the code.

Functions reduce computational effort by avoiding repetition and thus allow the programmer to reuse the previously written code whenever required. A function can accept one or more input values, process them, and return one or more output values.

In both MATLAB and Python, functions play a vital role in developing efficient programs. Functions can be used in the different ways depending upon the programming requirements:

- called from the main program,
- called from another function,
- stored in separate files, or
- defined directly using commands.

Before studying functions in MATLAB and Python separately, let us discuss the general layout and usage of a function.

## Structure of a function

A function generally contains the following components:

1. **Function definition**
   Specifies the name of the function and its inputs
2. **Function body**
   Contains the set of statements that perform the required task
3. **Return statement/output**
   Returns the computed result to the calling program

 | HTTPS://DOI.ORG/10.1515/9783112207758-007

**Syntax**

Function_Name(Input Arguments)

 Statements

 Processing

 Return Output

=Functions provide several advantages in programming by making programs more efficient, organized, and easy to manage. Moreover, functions help in reducing computational complexity and save development time by avoiding unnecessary rewriting of code.

## Calling a function

A function can be called in many different ways:
- from the main program,
- from another function, or
- directly through commands.

When a function is called, the control of execution transfers to the function. After execution, the control returns to the calling program.

Having discussed the basic concept of functions, further this chapter has the discussion on functions in detail for both MATLAB and Python. Each section explains the details on defining functions, passing arguments, returning values, calling functions, and practical examples for implementation.

## Functions in MATLAB

MATLAB provides a powerful mechanism to create user-defined functions for performing specific computational tasks. A MATLAB function is written using the function keyword and is generally stored in a separate file having the same name as the function.

### Structure of a MATLAB function

A MATLAB function consists of:
- function declaration,
- input arguments,
- processing statements, and
- output arguments.

### Syntax of a MATLAB function

A MATLAB function is defined using the ***function*** keyword followed by the output arguments, function name, and input arguments. The general structure of a MATLAB function is given below:

```
function [output_arguments] = function_name(input_arguments)
    % Executable statements
    % Processing operations
end
```

A MATLAB function is defined using the function keyword, followed by the output variables, function name, and input arguments. The output arguments specify the values returned after execution, whereas the input arguments represent the data supplied to the function for performing the required computations. The function name acts as the identifier through which the function is invoked within a program or from the Command Window. The body of the function contains the executable MATLAB statements and processing operations that accomplish the intended task. The definition of the function is terminated using the end statement, which marks the conclusion of the function body.

Let us consider an example with function definition, calling and getting the output. Note that the function file name and function name must be the same.

**Example 1: Function to find the square of a number**
**Function file**

```
function y = square_num(x)
y = x^2;
end
```

Save the file with the name: square_num.m

**Calling the function**
The function can be called from the Command Window or another script file:

```
a = 5;

result = square_num(a)
disp(result)
```

**Output:**

25

**Here is the detailed explanation:**

The function square_num accepts an input value x. The statement y = x^2 computes the square of the number. The computed value is returned through the output variable y.

The function is called using the statement:

result = square_num(a)

## Types of functions in MATLAB

Functions in MATLAB can be classified into different categories based on their usage and implementation. The appropriate choice of the function is based on the requirement for solving computational problem.

The commonly used types of functions in MATLAB are:

built-in functions,
user-defined functions,
anonymous functions,
nested functions, and
local functions

### Built-in functions

MATLAB provides a large collection of predefined functions known as built-in functions. These functions are already available in MATLAB and can be directly used without creating them manually. These functions help to perform the different mathematical operations, matrix manipulations, plotting, statistical analysis, and many other computational tasks. Examples of built-in functions are detailed in Table 7.1.

TABLE 7.1 Built-in functions and their details.

| Function | Purpose |
|---|---|
| **sqrt()** | Finds the square root |
| **sin()** | Computes the sine value |
| **cos()** | Computes the cosine value |
| **length()** | Finds the length of an array |
| **sum()** | Computes the sum of elements |
| **max()** | Finds the maximum value |

### Example 2: Using built-in functions

```
x = 25;
a = sqrt(x)
b = sin(pi/2)
c = max([4 7 2 9])

disp(a)
disp(b)
disp(c)
```

**Output:**

```
5
1
9
```

**Explanation:**
sqrt(x) computes the square root of 25.
sin(pi/2) computes the sine value of $\frac{\pi}{2}$.
max([4 7 2 9]) returns the largest element from the array.

## User-defined functions

User-defined functions are the functions that are created by the programmer to perform a specific task according to the problem requirements. These functions can be written using the function keyword and need to be stored in separate. m files.

**Example 3: User-defined function for addition**
**Create function file with two numbers as input**

```
function c = add_num(a,b)
c = a + b;
end
```

**Save the file as:** add_num.m

The defined function can be called from the console or from any other editor file but it should be in the same folder:

```
x = 10;
y = 20;
result = add_num(x,y)
disp(result)
```

**Output:**

30

**Explanation:**

The function add_num accepts two input values, computes their addition and return the result through the output variable c.

## Anonymous functions in MATLAB

In MATLAB, an anonymous function is a function which can be defined without creating a separate file. These functions are useful for writing a short mathematical expression or to perform simple operations in a compact form.

An anonymous function is created using the @ symbol, which is called the function handle operator. The variables specified after the @ symbol act as the input arguments of the function.

### Syntax of an anonymous function

```
function_name = @(input_arguments) expression
```

In this syntax, function_name represents the handle of the anonymous function, while the expression defines the operation to be performed on the input arguments.

**Example 4: Anonymous function for squaring a number**

```
square_num = @(x) x^2;
result = square_num(6)
disp(result)
```

**Output:**
36
The statement:
square_num = @(x) x^2;
creates an anonymous function that computes the square of a given number. Here, x is the input argument and x^2 is the mathematical operation performed by the function.
The function is called using:
square_num(6)
which returns the square of 6.
Anonymous functions simplify the implementation of short mathematical expressions without creating a separate function file.

## Nested functions in MATLAB

A nested function in MATLAB is a function which is defined within another function. The function inside the main function is known as the nested function. Nested functions are useful when a function needs to access or share the variables of the parent function.

The variables defined in the parent function can be directly used inside the nested function without passing them explicitly as input arguments. This feature makes nested functions highly useful for complex computational tasks and large-scale programming applications.

### Syntax of a nested function

```
function main_function()
      % Statements of main function
      function nested_function()
      % Statements of nested function
   end
end
```

**Example 5: Nested function for addition**

```
function main_add()
a = 10;
b = 20;
result = add_num();
disp(result)
   function c = add_num()
      c = a + b;
```

```
    end
end
```

**Output:**
30

In this example, the function add_num() is defined inside the main function main_-add(). The nested function directly accesses the variables a and b defined in the parent function without requiring them as input parameters. To get the output, it needs to call the function as main_add from the console or from the editor file.
The statement:
c = a + b;
computes the addition of the two variables and returns the result to the main function. Nested functions improve data sharing between functions and reduce the need for passing multiple arguments repeatedly. They enhance program organization and are useful in applications where helper functions are required only within a specific main function.

## Local functions in MATLAB

Local functions are the functions that are defined within the same file after the main script or primary function. These functions are accessible only within that particular file and cannot be directly called from other files unless they are separately defined.

Local functions help to organize the large programs by placing related functions together in a single file.

### Syntax of local functions

```
% Main Script or Function
Statements:
function output = function_name(input_arguments)

    % Function statements

end
```

**Example 6: Local function for multiplication**

```
clc;
clear;
```

```
a = 5;
b = 4;
result = multiply_num(a,b);
disp(result)
function c = multiply_num(x,y)
c = x * y;
end
```

**Output:**

20

In this example, the local function multiply_num is written within the same file as the main script. The variables a and b are passed as input arguments to the function.

The statement:

c = x * y;

performs the multiplication operation and returns the result to the main script.

Local functions improve program organization by allowing related functions to remain within a single file.

**Passing arguments to functions in MATLAB**

Arguments are the values that are supplied to a function for performing computations or operations. In MATLAB, functions can accept one or more input arguments depending on the requirements of the problem.

The input arguments are specified inside the parentheses of the function definition and are used within the function body for processing.

**Syntax for passing arguments**

```
function output = function_name(input_arguments)

% Processing statements

end
```

The values are passed to the function while calling it.

**Example: Function with multiple input arguments**

**Function file**

```
function z = multiply_num(x,y)

z = x * y;
```

```
end
```

Save the file as:
multiply_num.m

**Calling the function**

```
a = 4;
b = 7;

result = multiply_num(a,b)

disp(result)
```

**Output:**
28

The function multiply_num accepts two input arguments x and y. These arguments receive the values passed during the function call.
The statement:
z = x * y;
performs the multiplication operation and stores the result in the output variable z.
The function is called using:
multiply_num(a,b)
where the values of a and b are passed to the function as input arguments.

## Multiple output arguments in MATLAB functions

MATLAB functions can also return more than one output value. Multiple output arguments are enclosed within square brackets.

**Syntax for multiple outputs**

```
function [output1, output2] = function_name(input_arguments)
% Statements

end
```

**Example: Function returning sum and product**
**Function file**

```
function [s,p] = operations(a,b)
s = a + b;
```

```
p = a * b;
end
```

Save the file as:
operations.m

**Calling the function**

```
num1 = 5;
num2 = 3;

[sum_result, product_result] = operations(num1,num2)

disp(sum_result)
disp(product_result)
```

**Output:**

```
8
15
```

The function operations return two output values: s stores the sum of the input numbers, and p stores their product.
The outputs are received using:
[sum_result, product_result] = operations(num1,num2)
where the first returned value is assigned to sum_result and the second to product_result.

## Variable scope in MATLAB functions

The scope of a variable defines the region of a program where the variable can be accessed or used. In MATLAB, variables inside functions may behave differently depending on where they are declared. In MATLAB, variables are mainly classified into local and global variables.

**Local variables**
Local variables are variables that are defined within a function. These variables can only be accessed inside that particular function and are destroyed once the function execution is completed.

**Example: Local variable**

```
function result = local_example()
a = 10;
```

```
b = 20;
result = a + b;
end
```

In this example, the variables a and b are local variables because they are defined inside the function local_example. These variables cannot be accessed outside the function.

**Global variables**

Global variables are variables that can be accessed from multiple functions or scripts. To use a global variable in MATLAB, the keyword global must be declared in every function or script where the variable is required.

Global variables are useful when several functions need to share common data.

**Syntax of global variable**

```
global variable_name
```

**Example: Using global variables**

```
function global_example()
global x
x = 100;
display_value()
disp(x)
end

function display_value()
global x
x = x + 50;
end
```

**Output:**

```
150
```

In this example, the variable x is declared as a global variable using the global keyword. The value of x is shared between the functions global_example and display_value.

Initially:

```
x = 100;
```

The function display_value() increases the value of x by 50. Since x is global, the updated value becomes available throughout the program.

Proper management of variable scope improves program reliability and prevents unnecessary modification of variables. Local variables enhance security and modularity, whereas global variables simplify data sharing between multiple functions.

## Functions with default arguments in MATLAB

In MATLAB, functions can be designed to work with a variable number of input arguments. Sometimes, the programmer may not provide all the required inputs while calling a function. In such situations, default values can be assigned to the missing arguments within the function definition.

The nargin command in MATLAB is commonly used to determine the number of input arguments passed to a function. Based on the value of nargin, suitable default values can be assigned.

This approach improves the flexibility and usability of functions.

**Syntax using nargin**

```
function output = function_name(input_arguments)
if nargin < required_number
   % Assign default values
end
% Processing statements
end
```

**Example: Function with default argument**

**Function file**

```
function area = circle_area(r,pi_value)
if nargin < 2
   pi_value = pi;
end
area = pi_value * r^2;
end
```

Save the file as:

circle_area.m

**Calling the function**

**Case 1: Using a default value**

```
result = circle_area(5)
disp(result)
```

**Output:**
78.5398

**Case 2: Providing a user-defined value**
result = circle_area(5,3.14)
disp(result)

**Output:**
78.5000

The function circle_area computes the area of a circle. It accepts two input arguments:
- r represents the radius and
- pi_value represents the value of $\pi$.

The statement:
if nargin < 2
checks whether the second argument has been supplied during the function call. If the second argument is not provided, MATLAB automatically assigns the built-in value of pi using:
pi_value = pi;
Thus, the function can work both with user-defined values and default values. Functions with default arguments improve flexibility and reduce the need for repeatedly specifying commonly used values. They simplify function calls and make programs more user-friendly. Such functions also improve adaptability in scientific and engineering computations where certain parameters frequently remain constant.

## Functions without input and output arguments

In MATLAB, a function may also be created without using input arguments or output arguments. Such functions are generally used for displaying messages, performing predefined tasks, or executing operations that do not require external data.

These functions execute independently when called.

**Syntax**
function function_name()

% Statements

end

**Example: Displaying a message**

```
function greet()
disp('Welcome to MATLAB Programming')
end
```

Save the file as:
greet.m

**Calling the function**

```
greet()
```

**Output:**

Welcome to MATLAB Programming

The function greet() neither accepts any input argument nor returns any output value. The function simply executes the disp() statement whenever it is called.

**Variable-length input arguments (varargin)**

In MATLAB, there are situations where the number of input arguments may vary during different function calls. MATLAB provides the varargin feature to handle such cases.

The term varargin stands for variable-length input arguments. It allows a function to accept any number of input values. The arguments passed through varargin are stored as a cell array.

**Syntax**

```
function output = function_name(varargin)

   % Statements

end
```

**Example: Addition of multiple numbers**

```
function total = add_all(varargin)
total = sum([varargin{:}]);
end
```

Save the file as:
add_all.m

**Calling the function**

```
result1 = add_all(2,4,6)
```

```
result2 = add_all(1,3,5,7,9)

disp(result1)
disp(result2)
```

**Output:**

```
12
25
```

The function add_all accepts a variable number of input arguments using varargin. The statement:

[varargin{:}]

converts the cell array elements into a numerical array, while the sum() function computes their addition. Thus, the function can work with any number of input values.

**Variable-length output arguments (varargout)**

MATLAB also allows functions to return a variable number of output arguments using varargout. This feature provides flexibility when different numbers of outputs are required in different situations. The outputs returned through varargout are stored in a cell array.

**Syntax**

```
function varargout = function_name(input_arguments)

    % Statements

end
```

**Example: Returning multiple outputs**

```
function varargout = operations(a,b)
varargout{1} = a + b;
varargout{2} = a * b;
varargout{3} = a – b;
end
```

Save the file as:

operations.m

**Calling the function**

```
[s,p,d] = operations(10,5)
```

```
disp(s)
disp(p)
disp(d)
```

**Output:**

```
15
50
5
```

The function stores different output values inside the cell array varargout. Each output is accessed using indexing such as:
varargout{1}
which represents the first output argument.
This feature is useful when functions need to return varying numbers of results.

**Passing functions as arguments**

In MATLAB, functions can also be passed as arguments to other functions using function handles. This feature is highly useful in numerical methods, optimization problems, integration, and graphical computations.

A function handle is created using the @ operator.

**Example: Passing a function to integral()**

```
f = @(x) x^2;
result = integral(f,0,2)
disp(result)
```

**Output:**

```
2.6667
```

The anonymous function:
f = @(x) x^2;
defines the mathematical function $x^2$.
The statement:
integral(f,0,2)
passes the function handle f to the built-in MATLAB function integral() for computing the definite integral over the interval [0, 2].

**Applications of functions in MATLAB**

Functions are widely used in MATLAB for solving scientific, engineering, and mathematical problems efficiently. They help in developing modular programs and simplify complex computational tasks.

Some common applications of functions in MATLAB include numerical computations, matrix operations, solving differential equations, optimization techniques, data analysis, graphical plotting and engineering modeling.

## Summary of MATLAB functions

In this section, the concept of functions in MATLAB has been discussed in detail. MATLAB functions help in developing modular, organized, and reusable programs by dividing complex computational tasks into smaller manageable units. The syntax and structure of MATLAB functions, including input arguments, output arguments, and function definitions, were explained with suitable examples.

Different categories of MATLAB functions such as built-in functions, user-defined functions, anonymous functions, nested functions, local functions, and subfunctions were introduced to demonstrate their applications in programming and computational analysis. Advanced concepts including recursive functions, function handles, variable scope, default arguments, and variable-length input and output arguments were also discussed.

## Functions in Python

Python also provides a powerful and flexible mechanism for creating functions to perform specific computational tasks. Similar to MATLAB, functions in Python improve modularity, readability, and code reusability. However, Python functions are defined using the ***def*** keyword and follow indentation-based syntax.

Unlike MATLAB, Python does not require separate files for defining simple functions, although functions can also be stored in separate modules for large-scale programming applications.

In Python, functions can accept one or more input arguments, return one or more output values, call other functions, and support recursive as well as anonymous function definitions.

The following sections discuss different types and implementations of functions in Python with suitable examples.

### Structure of a Python function

A Python function is defined using the def keyword followed by the function name and input arguments enclosed within parentheses. The executable statements of the function are written inside an indented block.

### Syntax of a Python function

```
def function_name(input_arguments):
# Statements
# Processing operations
return output
```

A Python function begins with the def keyword, which is used to define the function. The function name acts as the identifier through which the function is called within the program. The input arguments specified inside the parentheses provide the data required for processing. The body of the function contains the executable Python statements written with proper indentation. The return statement is used to send the computed result back to the calling program.

**Example 1: Function to find the square of a number**

```
def square_num(x):
    y = x**2
    return y
a = 5
result = square_num(a)
print(result)
```

**Output:**

25

**Explanation:**

The function square_num() accepts an input value x and computes its square using the expression:

x**2

The computed result is returned using the return statement. The function is called using:

square_num(a)

where the value of a is passed as the input argument.

### Calling functions and passing arguments in Python

A Python function is executed by calling the function name followed by parentheses. The required input values are passed inside the parentheses as arguments.

Python functions can accept single or multiple input arguments and can also return one or more output values.

The return statement is used to send the computed result back to the calling program.

**Example 2: Function with multiple arguments**

```
def add_num(a,b):
    c = a + b
    return c

x = 10
y = 20
result = add_num(x,y)
print(result)
```

**Output:**

30

**Explanation:**

The function add_num() accepts two input arguments a and b. The statement:

c = a + b

computes the addition of the two numbers and stores the result in c. The computed value is returned using the return statement.

The function is called using:

add_num(x,y)

where the values of x and y are passed as input arguments.

## Functions returning multiple values

Python functions can also return multiple output values simultaneously. The returned values are separated by commas and are received using multiple variables.

**Syntax**

```
return value1, value2
```

**Example 3: Functions returning the sum and product**

```
def operations(a,b):
s = a + b
```

```
p = a * b
return s,p

num1 = 5
num2 = 3
sum_result, product_result = operations(num1,num2)
print(sum_result)
print(product_result)
```

**Output:**

```
8
15
```

**Explanation:**

The function operations() returns two output values: s stores the sum of the numbers, while p stores their product.

The returned values are received using:

sum_result, product_result = operations(num1,num2)

where the first returned value is assigned to sum_result and the second to product_result.

### Functions without arguments and return values

Python also allows functions that neither accept input arguments nor return output values. Such functions are generally used for displaying messages or performing predefined tasks.

**Example 4: Displaying a message**

```
def greet():
    print("Welcome to Python Programming")

greet()
```

**Output:**

Welcome to Python Programming

The function greet() does not require any input argument and does not return any output value. The function simply executes the print() statement whenever it is called.

### Default and keyword arguments in Python

Python allows functions to be defined with default argument values. If the user does not provide a value for a particular argument during the function call, the default value is automatically used.

Python also supports keyword arguments, where the arguments are passed using the parameter names. This improves readability and flexibility during function calls.

**Example 5: Functions with default argument**

```
def circle_area(r, pi_value = 3.14159):
    area = pi_value * r**2
    return area
result1 = circle_area(5)
result2 = circle_area(5,3.14)
print(result1)
print(result2)
```

**Output:**

```
78.53975
78.5
```

**Explanation:**

The function circle_area() accepts two input arguments:

- r represents the radius,
- pi_value represents the value of $\pi$.

The argument pi_value = 3.14159 acts as the default value. If the second argument is not supplied during the function call, Python automatically uses the default value.
The function call:

circle_area(5)

uses the default value of pi_value, whereas:

circle_area(5,3.14)

uses the user-defined value.

**Example 6: Keyword arguments**

```
def student(name, age):
    print("Name:", name)
    print("Age:", age)

student(age = 18, name = "John")
```

**Output:**

Name: John
Age: 18

**Explanation:**
In keyword arguments, the values are passed using the parameter names. The order of arguments does not matter when keyword arguments are used.

### Variable-length arguments in Python

Python allows functions to accept a variable number of arguments using:
– *args
– **kwargs
The *args parameter is used to pass multiple non-keyword arguments, whereas **kwargs is used to pass multiple keyword arguments.

**Example 7: Using *args**
```
def add_all(*args):
    total = sum(args)
    return total

result = add_all(2,4,6,8)
print(result)
```

**Output:**
20

**Explanation:**
The parameter *args allows the function to accept any number of input values. The values passed through args are stored as a tuple.
The statement:
sum(args)
computes the addition of all input values.

**Example 8: Using **kwargs**
```
def student_info(**kwargs):
    for key, value in kwargs.items():
        print(key, ":", value)

student_info(name = "John", age = 18, city = "London")
```

**Output:**

```
name: John
age: 18
city: London
```

**Explanation:**
The parameter **kwargs allows the function to accept multiple keyword arguments. The values are stored in the form of a dictionary.
The method:
kwargs.items()
returns the key-value pairs stored in the dictionary.

## Lambda and recursive functions in Python

Python supports anonymous functions known as lambda functions. These functions are useful for short mathematical operations and compact computations.

Python also supports recursive functions in which a function repeatedly calls itself until a terminating condition is satisfied.

**Example 9: Lambda function**

```
square_num = lambda x: x**2
result = square_num(5)
print(result)
```

**Output:**

```
25
```

**Explanation:**
The lambda function:
lambda x: x**2
computes the square of a number. Lambda functions are generally used for compact operations where defining a complete function is unnecessary.

**Example 10: Recursive function for factorial**

```
def factorial_num(n):
    if n == 0 or n == 1:
        return 1
    else:

        return n * factorial_num(n-1)

result = factorial_num(5)
```

```
print(result)
```

**Output:**

```
120
```

**Explanation:**
The function factorial_num() repeatedly calls itself using:
factorial_num(n-1)
until the terminating condition:
n == 0 or n == 1
is satisfied.
For $n = 5$:

$$5! = 5 \times 4 \times 3 \times 2 \times 1 = 120$$

## Variable scope in Python

In Python, variables defined inside a function are called local variables and can only be accessed within that function. Variables defined outside functions are generally treated as global variables.

Proper management of variable scope improves program reliability and simplifies debugging.

**Example 11: Local variable**

```
def local_example():
    x = 10
    print(x)
local_example()
```

**Output:**

```
10
```

**Example 12: Global variable**

```
x = 100
def global_example():
    global x
    x = x + 50
    print(x)
global_example()
```

**Output:**

150

**Explanation:**

In the first example, the variable x is local to the function local_example(). In the second example, the keyword global allows the variable x to be modified inside the function.

**Applications of functions in Python**

Functions in Python are widely used in scientific computing, data analysis, machine learning, web development, automation, and software development. They help in simplifying large programs, improving code reusability, and reducing computational complexity.

**Summary of Python functions**

In this section, the implementation of functions in Python has been discussed with suitable examples. Python functions provide an efficient way to organize programs into reusable and modular components. Different concepts including function definition, function calling, passing arguments, multiple return values, default arguments, keyword arguments, variable-length arguments, lambda functions, recursive functions, and variable scope were explained.

The section further demonstrated how Python functions simplify programming tasks, improve readability, and support efficient computational problem-solving in scientific and real-world applications.

## Exercise

### Section A: MATLAB functions

1. Write a MATLAB function to compute the cube of a given number.
2. Create a user-defined function to find the maximum of two numbers.
3. Write a MATLAB function that accepts two numbers and returns their sum, difference, and product.
4. Create an anonymous function to compute:

$$f(x) = x^2 + 3x + 5$$

5. Write a recursive MATLAB function to compute the factorial of a number.

6. Write a MATLAB function using varargin to compute the sum of multiple numbers.
7. Create a function with default arguments to compute the area of a circle.
8. Write a MATLAB program demonstrating the use of global variables.
9. Create a subfunction program to perform addition and multiplication of two numbers.
10. Write a MATLAB function that accepts another function as an argument and evaluates it over a given interval.

### Section B: Python functions

1. Write a Python function to compute the square of a number.
2. Create a Python function to find the average of three numbers.
3. Write a Python function that returns the sum and product of two numbers.
4. Create a lambda function to compute:

$$f(x) = x^3 + 2x + 1$$

5. Write a recursive Python function to compute the Fibonacci series.
6. Create a Python function using default arguments to calculate simple interest.
7. Write a Python program using keyword arguments to display student information.
8. Create a Python function using *args to compute the product of multiple numbers.
9. Write a Python program demonstrating the use of global variables.
10. Create a Python function that accepts another function as an argument and evaluates it for different input values.

## Computational practice questions with short answers

1. **Evaluate a quadratic function**
   **Question**
   Write a function to evaluate:

$$f(x) = x^2 + 3x + 2$$

   Find $f(5)$.

**MATLAB code**

**Python code**

**Answer:** 42

2. **Area of a circle**
   **Question**
   Calculate the area of a circle for radius $r$ = 7.

$$A = \pi r^2$$

**MATLAB code**

```
function y = quadratic(x)
y = x^2 + 3*x + 2;
end
quadratic(5)
```

```
def quadratic(x):
return x**2 + 3*x + 2
print(quadratic(5))
```

**Python code**

**Answer:** 153.94

3. **Distance between two points**
   **Question**
   Find the distance between points (2,3) and (8,11).

**MATLAB code**

**Python code**

```
function A = area_circle(r)
A = pi*r^2;
end
area_circle(7)
```

**Answer:** 10

```
import math

def area_circle(r):
    return math.pi*r**2

print(area_circle(7))
```

4. **Simple interest**
   **Question**
   Calculate simple interest for:
   - $P$ = 5,000
   - $R$ = 8%
   - $T$ = 3 years

**MATLAB code**

```
function d = distance(x1,y1,x2,y2)
    d = sqrt((x2-x1)^2 + (y2-y1)^2);
end

distance(2,3,8,11)
```

```
import math
def distance(x1,y1,x2,y2):

    return math.sqrt((x2-x1)**2 + (y2-y1)**2)

print(distance(2,3,8,11))
```

**Python code**

**Answer:** 1200

5. **Hypotenuse of a right triangle**
   **Question**
   Using the Pythagoras theorem, find the hypotenuse when:
   - $a$ = 5

- $b = 12$

```
function SI = simple_interest(P,R,T)
    SI = (P*R*T)/100;
end

simple_interest(5000,8,3)
```

**MATLAB code**

```
def simple_interest(P,R,T):
    return (P*R*T)/100

print(simple_interest(5000,8,3))
```

**Python code**

**Answer:** 13

6. **Factorial using a function**
   **Question**
   Find 7!.

**MATLAB code**

```
function h = hypotenuse(a,b)
    h = sqrt(a^2+b^2);
end

hypotenuse(5,12)
```

```
import math

def hypotenuse(a,b):
    return math.sqrt(a**2+b**2)

print(hypotenuse(5,12))
```

### Python code

**Answer:** 5040

```
function f = factorial_fun(n)
    f = 1;
    for i = 1:n
        f = f*i;
    end
end

factorial_fun(7)
```

7. **Sum of squares**
   **Question**

```
def factorial_fun(n):
    f = 1
    for i in range(1,n+1):
        f *= i
    return f

print(factorial_fun(7))
```

Calculate:

$$1^2 + 2^2 + 3^2 + 4^2 + 5^2$$

### MATLAB code

### Python code

```
function s = sumSquares(n)
    s = 0;
    for i = 1:n
        s = s + i^2;
    end
end

sumSquares(5)
```

**Answer:** 55

```
def sumSquares(n):
    s = 0
    for i in range(1,n+1):
        s += i**2
    return s

print(sumSquares(5))
```

8. **Roots of a quadratic equation**
   **Question**
   Find the roots of:

$$x^2 - 5x + 6 = 0$$

### MATLAB code

```
function [r1,r2] = quadraticRoots(a,b,c)
    d = b^2 - 4*a*c;
    r1 = (-b + sqrt(d))/(2*a);
    r2 = (-b - sqrt(d))/(2*a);
end

[r1,r2] = quadraticRoots(1,-5,6)
```

### Python code

```
import math

def quadraticRoots(a,b,c):
    d = b**2 - 4*a*c
    r1 = (-b + math.sqrt(d))/(2*a)
    r2 = (-b - math.sqrt(d))/(2*a)
    return r1,r2

print(quadraticRoots(1,-5,6))
```

**Answer:** 3 and 2

9. **Determinant of a matrix**
   **Question**
   Find the determinant of

$$\begin{bmatrix} 2 & 1 \\ 3 & 4 \end{bmatrix}$$

**MATLAB code**

```
A = [2 3;
     1 4];

det(A)
```

**Python code**

```
import numpy as np

A = np.array([[2,3],
   [1,4]])

   print(np.linalg.det(A))
```

**Answer:** 5

10. **Numerical integration**
    **Question**
    Evaluate:

$$\int_{0}^{3} x^2 dx$$

**MATLAB code**

```
f = @(x) x.^2;
integral(f,0,2)
```

**Python code**

```
from scipy.integrate import quad

def f(x):
    return x**2

result, error = quad(f,0,2)
print(result)
```

**Answer:** 2.6667

11. **Compound interest**
    **Question**
    Calculate the compound amount:

$$A = P\left(1 + \frac{r}{100}\right)^t$$

for $P = 10000$, $r = 10\%$, $t = 2$.

**MATLAB code**

```
function A = compoundInterest(P,r,t)
    A = P*(1+r/100)^t;
end

compoundInterest(10000,10,2)
```

**Python code**

```
def compoundInterest(P,r,t):
    return P*(1+r/100)**t

print(compoundInterest(10000,10,2))
```

**Answer:** 12100

12. **Arithmetic progression sum**
    **Question**
    Find the sum of first 10 terms of the AP:
    2, 4, 6, ...

### MATLAB code

```
function S = APSum(a,d,n)
    S = n/2*(2*a + (n-1)*d);
end

APSum(2,2,10)
```

### Python code

```
def APSum(a,d,n):
    return n/2*(2*a + (n-1)*d)

print(APSum(2,2,10))
```

**Answer:** 110

13. **Geometric progression sum**
    **Question**
    Find the sum of first five terms of GP:
    2, 4, 8, 16, 32

### MATLAB code

```
function S = GPSum(a,r,n)
    S = a*(r^n - 1)/(r - 1);
end

GPSum(2,2,5)
```

### Python code

```
def GPSum(a,r,n):
    return a*(r**n - 1)/(r - 1)

print(GPSum(2,2,5))
```

**Answer:** 62

14. **Evaluate exponential function**
    **Question**
    Find $e^2$.

**MATLAB code**

```
function y = exponentialValue(x)
    y = exp(x);
end

exponentialValue(2)
```

**Python code**

```
import math

def exponentialValue(x):
    return math.exp(x)

print(exponentialValue(2))
```

**Answer:** 7.389

15. **Logarithm base 10**
    **Question**
    Find:

$$\log_{10}(1000)$$

**MATLAB code**

```
function y = logValue(x)
    y = log10(x);
end

logValue(1000)
```

**Python code**

```
Import math

def logValue(x):
    return math.log10(x)

print(logValue(1000))
```

**Answer:** 3

16. **Sine function**
    **Question**
    Find:

$$\sin(30^\circ)$$

**MATLAB code**

```
function y = sineValue(theta)
    y = sind(theta);
end

sineValue(30)
```

**Python code**

```
import math

def sineValue(theta):
    return math.sin(math.radians(theta))

print(sineValue(30))
```

**Answer:** 0.5

17. **Verify trigonometric identity**
    **Question**
    Evaluate:

$$sin^2(45^\circ) + cos^2(45^\circ)$$

## MATLAB code

```
function y = trigIdentity()
    y = sind(45)^2 + cosd(45)^2;
end

trigIdentity()
```

## Python code

```
import math

def trigIdentity():
    return math.sin(math.radians(45))**2 +
        math.cos(math.radians(45))**2

print(trigIdentity())
```

**Answer:** 1

18. **Area of triangle (Heron's formula)**
    **Question**
    Find the area of a triangle with sides 5, 6, and 7.

## MATLAB code

```
function A = triangleArea(a,b,c)
    s = (a+b+c)/2;
    A = sqrt(s*(s-a)*(s-b)*(s-c));
end

triangleArea(5,6,7)
```

**Python code**

```
import math

def triangleArea(a,b,c):
    s = (a+b+c)/2
    return math.sqrt(s*(s-a)*(s-b)*(s-c))

print(triangleArea(5,6,7))
```

**Answer:** 14.697

19. **Volume of a sphere**
    **Question**
    Find the volume of a sphere of radius 3.

**MATLAB code**

```
function V = sphereVolume(r)
    V = (4/3)*pi*r^3;
end

sphereVolume(3)
```

**Python code**

```
import math

def sphereVolume(r):
    return (4/3)*math.pi*r**3

print(sphereVolume(3))
```

**Answer:** 113.097

20. **Average of array elements**
    **Question**
    Find the average of:
    [10, 20, 30, 40, 50]

**MATLAB code**

```
function A = averageValue(x)
   A = mean(x);
end

averageValue([10 20 30 40 50])
```

**Python code**

```
def averageValue(x):
   return sum(x)/len(x)

print(averageValue([10,20,30,40,50]))
```

**Answer:** 30

21. **Standard deviation**
    **Question**
    Find the standard deviation of:
    [2, 4, 6, 8, 10]

**MATLAB code**

```
function s = stdValue(x)
   s = std(x,1);
end

stdValue([2 4 6 8 10])
```

**Python code**

```
import numpy as np

def stdValue(x):
   return np.std(x)

print(stdValue([2,4,6,8,10]))
```

**Answer:** 2.828

22. **Fibonacci number**
    **Question**
    Find the 10th Fibonacci number.

**MATLAB code**

```
function f = fibonacci(n)
    a = 0; b = 1;
    for i = 2:n
        c = a + b;
        a = b;
        b = c;
    end
    f = b;
end

fibonacci(10)
```

**Python code**

```
def fibonacci(n):
    a,b = 0,1
    for i in range(2,n+1):
        a,b = b,a+b
    return b

print(fibonacci(10))
```

**Answer:** 55

23. **Evaluate polynomial**
    **Question**
    Evaluate:

$$f(x) = 2x^3 - 5x^2 + 4x - 1$$

at $x = 3$.

### MATLAB code

```
function y = polynomial(x)
    y = 2*x^3 - 5*x^2 + 4*x - 1;
end

polynomial(3)
```

### Python code

```
def polynomial(x):
    return 2*x**3 - 5*x**2 + 4*x - 1

print(polynomial(3))
```

**Answer:** 26

24. **RMS value**
    **Question**
    Find the RMS value of:
    [2, 4, 6, 8]

### MATLAB code

```
function r = rmsValue(x)
    r = sqrt(mean(x.^2));
end

rmsValue([2 4 6 8])
```

### Python code

```
import numpy as np

def rmsValue(x):
    return np.sqrt(np.mean(np.array(x)**2))

print(rmsValue([2,4,6,8]))
```

**Answer:** 5.477

25. **Numerical differentiation**
**Question**
Find the derivative of:

$$f(x) = x^3$$

at $x = 2$ using

$$f'(x) = \frac{f(x+h) - f(x)}{h}$$

where $h = 0.001$.

### MATLAB code

```
function d = derivative(x)
    h = 0.001;
    f = @(x) x.^3;
    d = (f(x+h)-f(x))/h;
end

derivative(2)
```

### Python code

```
def derivative(x):
    h = 0.001
    f = lambda x: x**3
    return (f(x+h)-f(x))/h

print(derivative(2))
```

**Answer:** ≈ 12.006 (approximately 12)

## Mini project: mathematical function analyzer in MATLAB and Python

### Objective

Develop a program that uses user-defined functions to analyze mathematical functions and compute various properties such as:
- function evaluation,
- derivative approximation,
- definite integration,
- root finding,
- maximum and minimum values, and
- graph plotting

This project demonstrates the power of modular programming using functions in both MATLAB and Python.

### Problem statement

Create a system that analyzes the function:

$$f(x) = x^3 - 6x^2 + 11x - 6$$

The program should provide the following options:
1. Evaluate $f(x)$
2. Compute derivative at a point
3. Compute definite integral
4. Find roots of the equation
5. Find maximum and minimum values over an interval
6. Plot the function
7. Exit

Each operation must be implemented as a separate user-defined function.

CHAPTER 8

# Plotting and data visualization in MATLAB and Python

Visualization is important in scientific computing, engineering analysis, data interpretation, and mathematical modeling. Numerical data is easier to understand when it is represented graphically. Programmers can use plotting and data visualization techniques to analyze trends, compare datasets, identify patterns, and present computational results.

MATLAB and Python offer powerful tools and built-in functions to generate graphical representation of data in two-dimensional and three-dimensional forms. In addition to plotting numerical data, these programming environments also provide facilities to edit the produced graph for customizing the colors, style and the presentation.

MATLAB has an integrated plotting environment and special commands for plotting. Python has wide support for plotting through libraries such as Matplotlib, NumPy, and Seaborn. Both languages can produce high-quality graphical outputs for scientific research, engineering applications and data analysis.

This chapter deals with tools and techniques for plotting and data with a comparative understanding of graphical functions available in both the programming environments.

## Plotting and visualization in MATLAB

MATLAB provides a powerful graphical environment for visualizing numerical data and mathematical functions. It contains a large collection of built-in plotting commands that help programmers to generate a high-quality graphical representation.

MATLAB plots are highly interactive and customizable. A programmer can modify line styles, colors, axes, labels, legends, titles, grids, and dimensions according to the requirements of the problem. MATLAB also supports both two-dimensional and three-dimensional plotting techniques for representing simple as well as multidimensional datasets

## Basic 2D plot in MATLAB

The most commonly used plotting function in MATLAB is the **plot()** function. It is used to create two-dimensional line graphs between two sets of data.

 | HTTPS://DOI.ORG/10.1515/9783112207758-008

The plot() command takes the values of the horizontal axis and vertical axis as input arguments and generates the corresponding graph.

**Syntax**

```
plot(x,y)
```

where x represents the values along the horizontal axis and y represents the values along the vertical axis.

**Example 1: Plotting a simple line graph**

```
x=0:1:10;>
y=x.^2;>
plot(x,y)>
grid on
```

**Explanation**

In this example, x=0:1:10 generates the values from 0 to 10 with increment 1, y = x.^2 computes the square of each value of x, plot(x,y) generates the graph between x and y, and grid on displays the grid lines on the graph.

The resulting graph represents the quadratic function:

The graph appears as a smooth upward curving parabola, as shown in Figure 8.1.

The plot() function supports multiple graph plotting, line color modification, marker customization, line style formatting, axis labeling, title insertion, and graphical annotations. These customization features help improve the readability and presentation quality of graphical outputs.

## Customizing 2D plots in MATLAB

MATLAB provides several options for customizing graphical outputs to improve clarity, readability, and presentation quality.

### Adding title and axis labels

Titles and labels provide proper description and interpretation of the graph. MATLAB uses the following commands:

- title() for adding graph title,
- xlabel() for labeling the horizontal axis, and
- ylabel() for labeling the vertical axis.

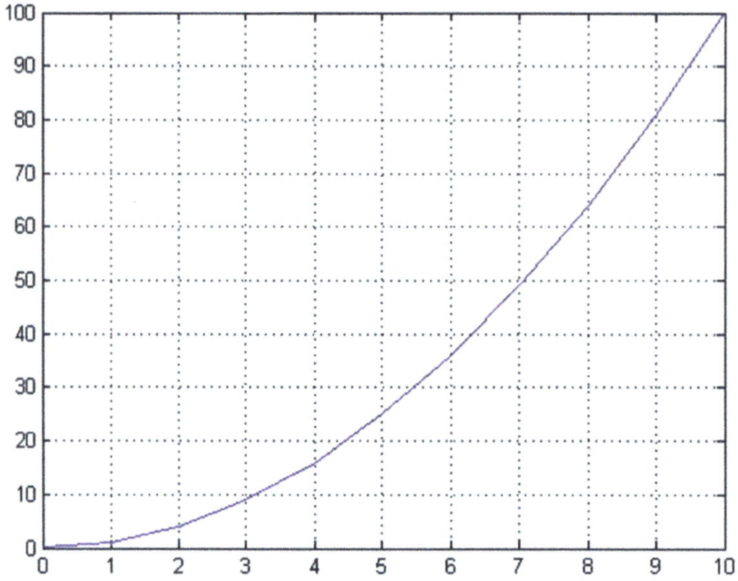


FIGURE 8.1 The graph of parabola appeared from commands.

**Example: Plot with labels and title**

```
x=0:1:10;
y=x.^2;
plot(x,y)
title('Graph of y=x^2')
xlabel('X-axis')
ylabel('Y-axis')
grid on
```

**Explanation**

In this example, title() adds the title to the graph, xlabel() labels the horizontal axis, and ylabel() labels the vertical axis, grid on displays grid lines for better visualization. These commands improve the presentation and readability of the graphical output, as shown in Figure 8.2.

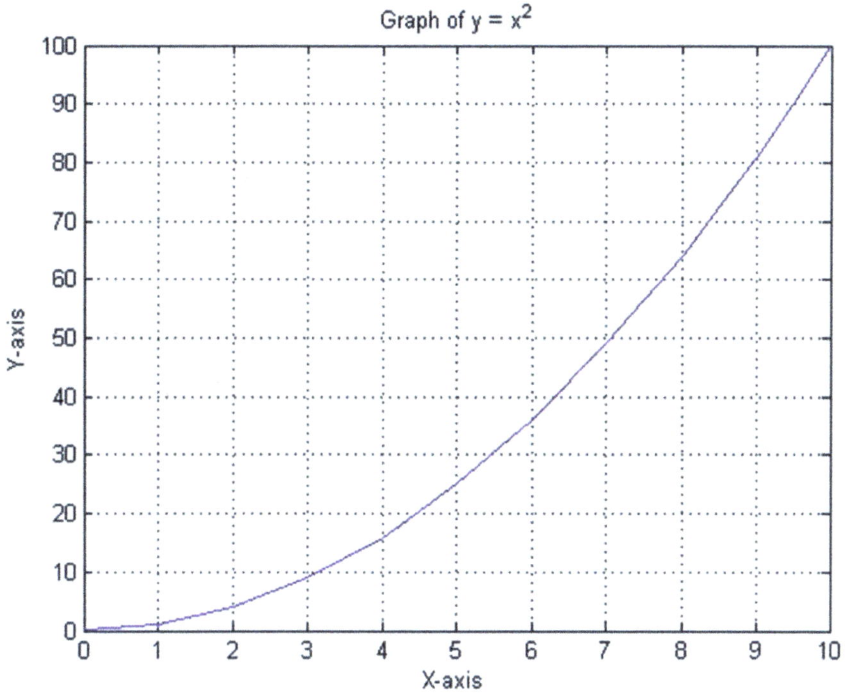


FIGURE 8.2 Updated graph of the curve with added axis labels.

### Modifying line color and style

MATLAB allows the programmer to modify line appearance using color codes, line styles, and markers directly within the plot() command.

The details of the common color codes are tabulated in Table 8.1 with the common line styles mentioned in Table 8.2 and the common marker styles in Table 8.3.

TABLE 8.1 Color coding of graphs.

| Code | Color |
|---|---|
| 'r' | Red |
| 'g' | Green |
| 'b' | Blue |
| 'k' | Black |
| 'm' | Magenta |
| 'c' | Cyan |

TABLE 8.2 Common line styles of graphs.

| Symbol | Style |
|---|---|
| '-' | Solid line |
| '--' | Dashed line |
| ':' | Dotted line |
| '-.' | Dash-dot line |

TABLE 8.3 Common markers of graphs.

| Symbol | Marker |
|---|---|
| 'o' | Circle |
| '*' | Star |
| '+' | Plus |
| 'x' | Cross |
| 's' | Square |

**Example: Customized plot**

```
x=0:1:10;
y=x.^2;
plot(x,y,'r–o','LineWidth',2)
title('Customized Plot')
xlabel('X-axis')
ylabel('Y-axis')
grid on
```

**Explanation**

In the plotting command, r specifies red color, – specifies dashed line style, and o specifies circular markers.

The command 'LineWidth',2 increases the thickness of the graph line. The resulting graph appears as a red dashed curve with circular markers. The output is displayed in Figure 8.3.

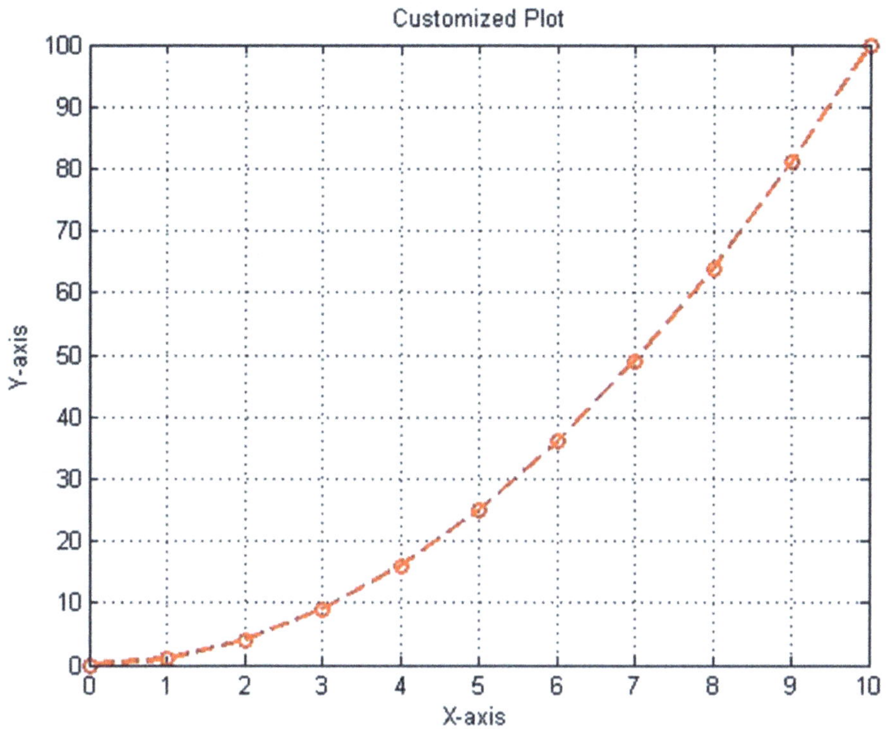


FIGURE 8.3 Modified graph with markers and line styles.

### Plotting multiple curves on the same graph

MATLAB allows multiple datasets to be plotted within the same figure window. This feature is useful for comparing different mathematical functions or datasets. Multiple curves can be plotted by passing additional sets of coordinates inside the plot() command.

**Example: Plotting multiple curves**

```
x = −10:0.1:10;
y1 = x.^2;
y2 = x.^3;
plot(x,y1,'b',x,y2,'r–')
legend('y = x^2','y = x^3')
title('Multiple Curves')
```

xlabel('X-axis')
ylabel('Y-axis')
grid on

**Explanation**

In this example, the first curve represents $y = x^2$, the second curve represents $y = x^3$, legend() identifies individual curves, and different colors and line styles distinguish the graphs visually. The graph displays both functions simultaneously within the same coordinate system for comparison. The output is shown in Figure 8.4.

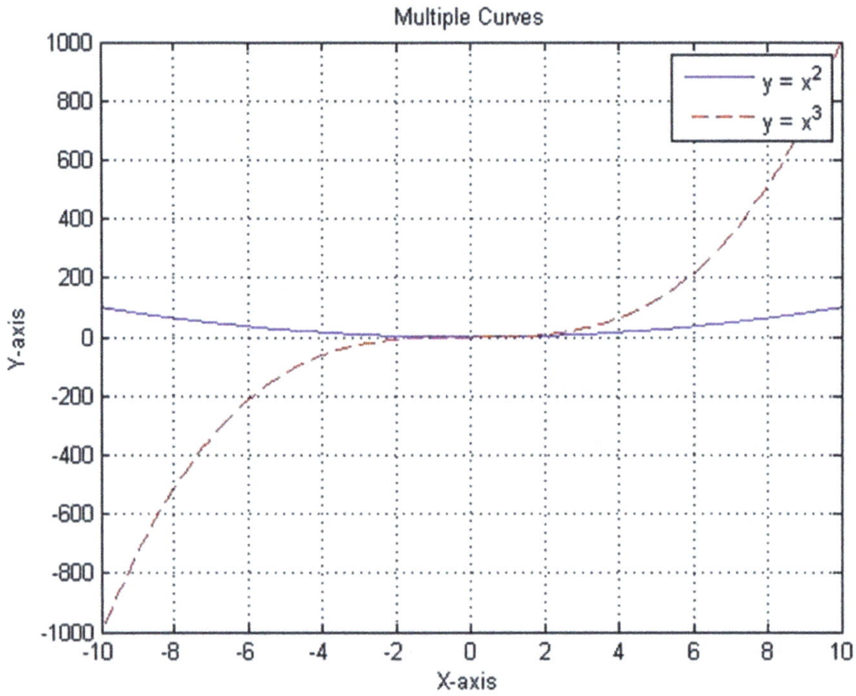


FIGURE 8.4 Plot of multiple graphs in a single figure window.

## Subplots in MATLAB

In many scientific and engineering applications, it is often necessary to display multiple graphs within the same figure window for comparison and analysis. MATLAB provides the subplot() function to divide the figure window into multiple smaller plotting regions.

Each subplot acts as an independent plotting area where separate graphs can be displayed.

**Syntax of subplot()**

```
subplot(m,n,p)
```

where **m** represents the number of rows, **n** represents the number of columns, and **p** represents the position of the current plot.

**Example: Creating multiple subplots**

```
x = −10:0.1:10;
y1 = sin(x);
y2 = cos(x);
subplot(2,1,1)
plot(x,y1)
title('Sine Function')
grid on
subplot(2,1,2)
plot(x,y2)
title('Cosine Function')
grid on
```

**Explanation**

In this example:

- subplot(2,1,1) divides the figure into two rows and one column and activates the first plotting region and
- subplot(2,1,2) activates the second plotting region.

The first subplot displays the sine function, whereas the second subplot displays the cosine function. Subplots help compare multiple datasets or functions simultaneously within a single figure window. The output is shown in Figure 8.5.

## Scatter plots in MATLAB

A scatter plot is used to represent individual data points on a graph. Scatter plots are widely used in statistics, data analysis, machine learning, and pattern recognition to study relationships between variables.

MATLAB provides the scatter() function for generating scatter plots.

**Syntax**

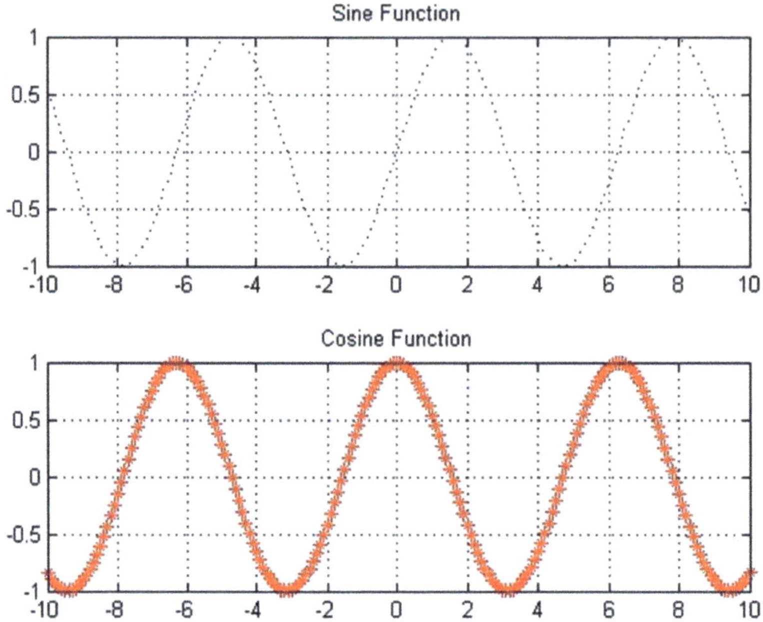


FIGURE 8.5 Demonstration of the subplots in a figure window.

scatter(x,y)

**Example: Scatter plot**

```
x = [1 2 3 4 5 6];
y = [2 4 5 4 6 7];
scatter(x,y,'filled')
title('Scatter Plot')
xlabel('X-axis')
ylabel('Y-axis')
grid on
```

**Explanation**

In this example scatter(x,y) plots, the individual data points, the option 'filled' fills the markers with color, each point represents a pair of values from x and y. The output is shown in Figure 8.6.

Scatter plots are useful for identifying trends, clustering patterns, and relationships between variables.

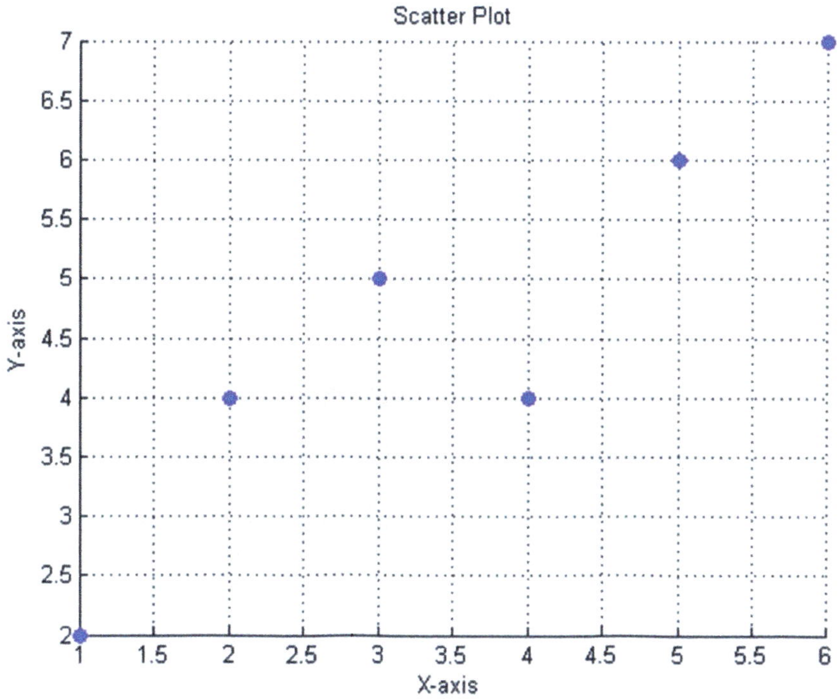


FIGURE 8.6 Demonstration of the scatter plot for a finite dataset.

## Bar graphs in MATLAB

Bar graphs are used for representing categorical or discrete data using rectangular bars. The height of each bar corresponds to the magnitude of the data value.

MATLAB provides the bar() function to generate bar graphs.

### Syntax

```
bar(x,y)
```

### Example: Bar graph

```
x = [1 2 3 4 5];
y = [5 7 3 8 6];
bar(x,y)
title('Bar Graph')
```

```
xlabel('Category')
ylabel('Values')
grid on
```

**Explanation**
In this example, the values stored in x represent categories, the values stored in y determine the heights of the bars, bar(x,y) generates the bar graph. See the output in Figure 8.7.

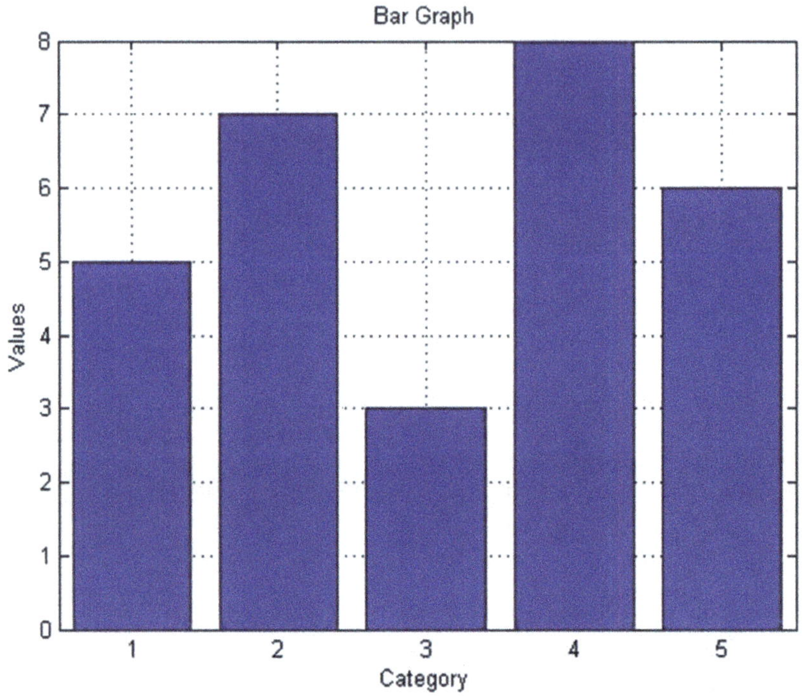


FIGURE 8.7 Bar graph of the discrete data points.

Bar graphs are commonly used for comparison of categorical datasets.

## Histograms in MATLAB

Histograms are graphical representations used to visualize the frequency distribution of numerical data. They divide the data into intervals called bins and display the frequency of values within each interval.

MATLAB provides the histogram() function for creating histograms.

**Syntax**

```
histogram(data)
```

**Example: Histogram plot**

```
data = randn(1,1000);
hist(data)
title('Histogram')
xlabel('Values')
ylabel('Frequency')
grid on
```

**Explanation**

In this example, randn(1,1,000) generates 1,000 random numbers with normal distribution, histogram(data) groups the data into bins and plots the frequency distribution, as shown in Figure 8.8.

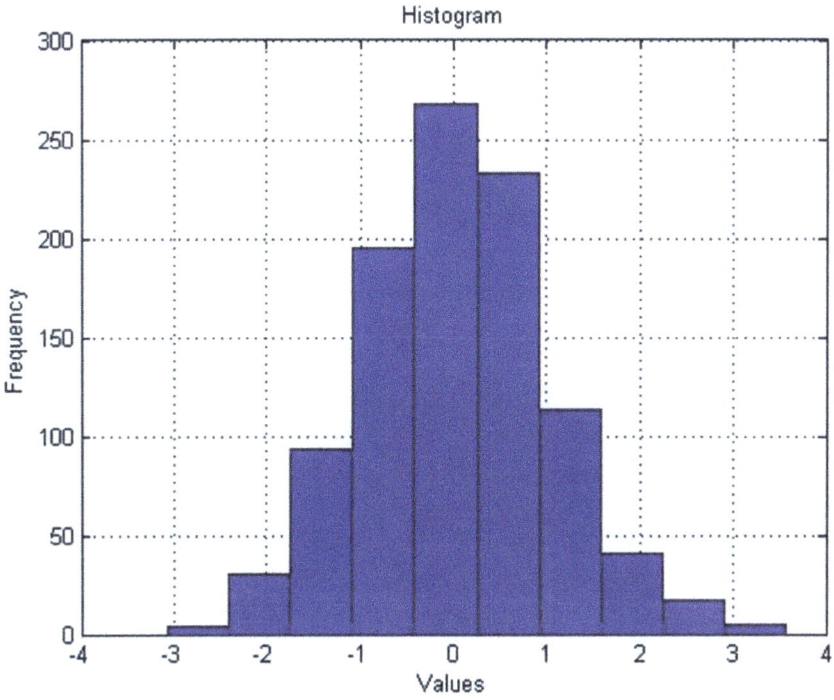


FIGURE 8.8 Histogram of the random data values classified in bins.

Histograms are widely used in statistical analysis and probability studies to understand data distribution patterns.

## Pie charts in MATLAB

Pie charts are circular graphical representations used to display the proportional distribution of data. Each sector of the pie represents a fraction or percentage of the total dataset.

Pie charts are widely used in statistics, business analysis, surveys, and data comparison applications.

MATLAB provides the pie() function for generating pie charts.

**Syntax**

```
pie(data)
```

**Example: Pie chart**

```
data = [25 15 30 10 20];
pie(data)
title('Pie Chart Representation')
```

**Explanation**

In this example, the vector data contains the numerical values, pie(data) divides the circular graph into sectors proportional to the data values, as shown in Figure 8.9.

Each sector represents the relative contribution of the corresponding data value to the total dataset.

Labels can be added to identify different sectors of the pie chart.

**Example:**

```
data = [20 30 25 25];
labels = {'A','B','C','D'};
pie(data,labels)
title('Pie Chart with Labels')
```

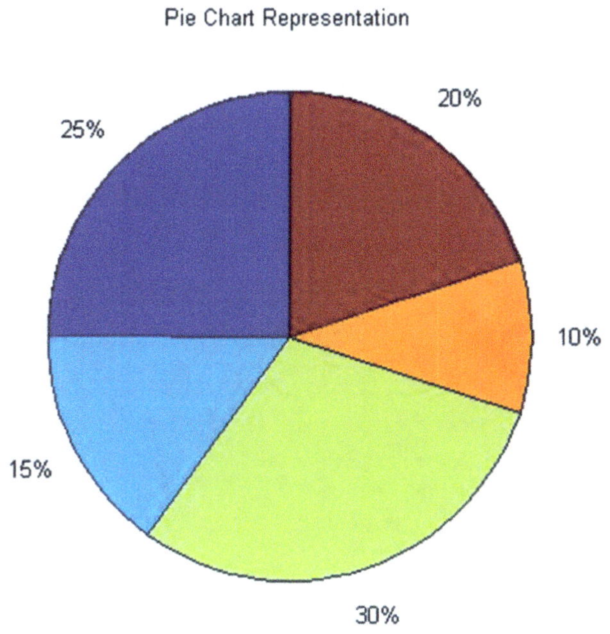


FIGURE 8.9 Pie chart of the data value showing the sector distributions.

**Explanation**
The cell array {'A','B','C','D'} contains labels corresponding to each sector of the pie chart. These labels improve readability and interpretation of the graphical representation.

## 3D plotting in MATLAB

MATLAB also provides powerful tools for generating three-dimensional graphical representations. Three-dimensional plots are useful for visualizing surfaces, multidimensional functions, engineering simulations, and scientific computations.

Commonly used 3D plotting functions in MATLAB include plot3(), mesh(), surf().

### 3D line plot using plot3()

The plot3() function is used to generate three-dimensional line plots.

**Syntax**

```
plot3(x,y,z)
```

**Example: 3D line plot**

```
t = -10:0.1:10;
x = sin(t);
y = cos(t);
z = t;
plot3(x,y,z)
grid on
title('3D Line Plot')
xlabel('X-axis')
ylabel('Y-axis')
zlabel('Z-axis')
```

**Explanation**

In this example, x = sin(t) and y = cos(t) generates circular coordinates, z = t provides height variation, and plot3(x,y,z) generates the three-dimensional spiral curve. The outcome is shown in Figure 8.10.

The commands xlabel(), ylabel(), and zlabel() label the three coordinate axes.

### Surface plots using surf

Surface plots are used to visualize functions of two variables in three-dimensional form. MATLAB provides the surf() function for generating colored surface plots.

**Syntax**

```
surf(X,Y,Z)
```

**Example: Surface plot**

```
[X,Y] = meshgrid(-5:0.5:5);
Z = X.^2 + Y.^2;
surf(X,Y,Z)
```

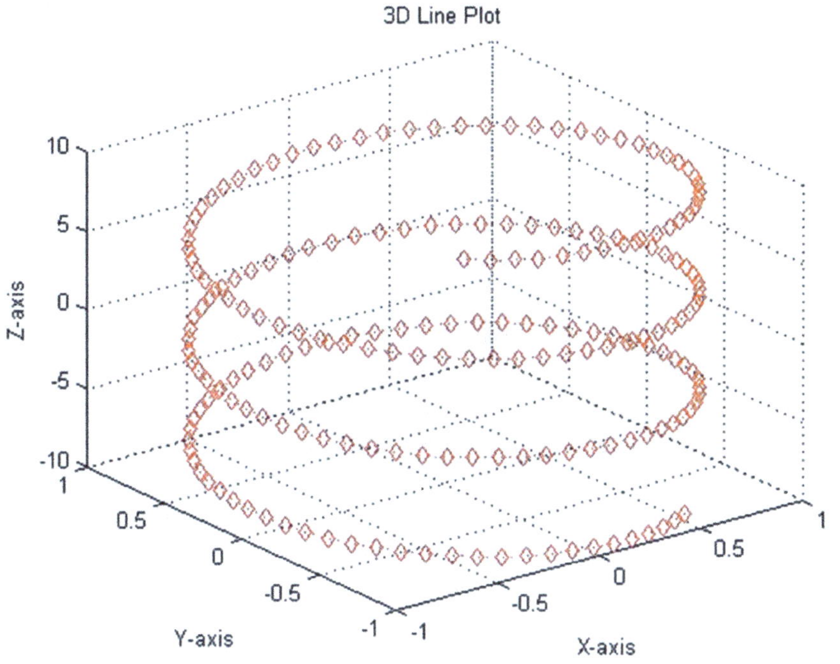


FIGURE 8.10 Demonstration of the 3D line plot.

```
title('Surface Plot')
xlabel('X-axis')
ylabel('Y-axis')
zlabel('Z-axis')
```

**Explanation**

In this example, meshgrid() generates coordinate matrices, Z = X.^2 + Y.^2 computes the surface values, surf() generates the three-dimensional colored surface plot. The graph is shown in Figure 8.11.

The resulting graph represents a paraboloid surface. Surface plots are extensively used in mathematical modeling, optimization, and engineering analysis.

### Mesh plots in MATLAB

A mesh plot represents a three-dimensional surface using a wireframe structure. MATLAB provides the mesh() function for generating mesh plots.

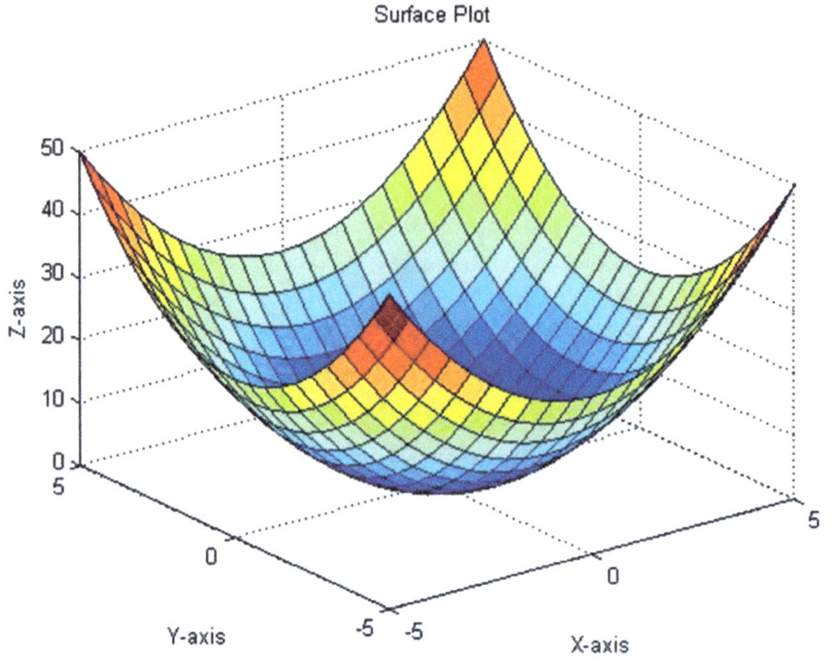


FIGURE 8.11 Presentation of a surface plot using meshgrid command.

**Example: Mesh plot**

```
[X,Y] = meshgrid(-5:0.5:5);
Z = sin(sqrt(X.^2 + Y.^2));
mesh(X,Y,Z)
title('Mesh Plot')
xlabel('X-axis')
ylabel('Y-axis')
zlabel('Z-axis')
```

The mesh() function creates a wireframe representation of the surface defined by the matrix Z. See Figure 8.12 for the output. Mesh plots are useful for studying surface structures and variations in three-dimensional space.

**Advantages of MATLAB visualization tools**

MATLAB visualization tools provide powerful capabilities for representing numerical and multidimensional data graphically. These tools simplify interpretation of complex datasets and improve analytical understanding. The built-in visualization environment of MATLAB allows programmers to generate high-quality graphical outputs efficiently with minimal programming effort:

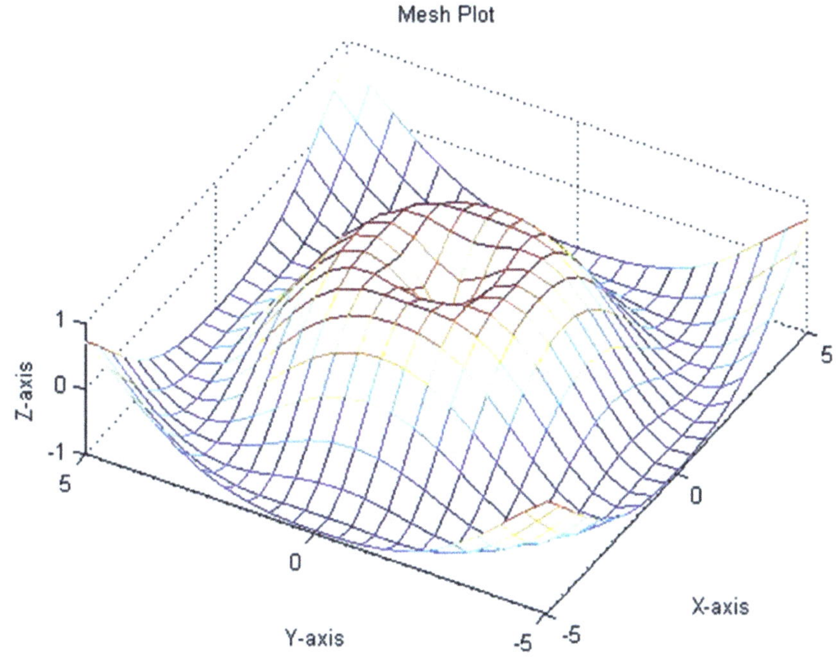


FIGURE 8.12 Generation of surface using mesh command.

Top of form
Bottom of form

## Saving and exporting plots in MATLAB

MATLAB allows graphical outputs to be saved in different image and figure formats for documentation, presentations, publications, and further analysis. The generated plots can be saved as a MATLAB figure files, image files, PDF documents, or high-resolution graphics.

Saving plots is important for preserving graphical results and using them in reports or research work.

**Saving a figure using saveas()**

The saveas() function in File option can be used to save the current figure in different formats such as PNG, JPG, PDF, and FIG.

**Saving figures using Export()**

MATLAB also provides the Export function in File option for exporting high-quality graphical outputs.

## Figure window customization in MATLAB

MATLAB provides various commands for customizing figure windows and improving graphical presentation.

Some commonly used commands are provided in Table 8.4.

TABLE 8.4 Commonly used MATLAB commands for figure customization

| Command | Purpose |
|---|---|
| **figure** | Opens a new figure window |
| **hold on** | Retains current graph while adding new plots |
| **hold off** | Releases the current graph |
| **axis()** | Controls axis limits |
| **grid on** | Displays grid lines |
| **legend()** | Adds legends to graphs |

**Example: Using hold on**

```
x = 0:0.1:10;
y1 = sin(x);
y2 = cos(x);
plot(x,y1)
hold on
plot(x,y2)
legend('sin(x)','cos(x)')
grid on
```

**Explanation**

In this example, hold on allows multiple graphs to appear in the same figure window, the sine and cosine curves are displayed together, legend() identifies individual graphs.

## Axis control in MATLAB

The axis() command is used to control the plotting range and scaling of graphs.

**Syntax**

```
axis([xmin xmax ymin ymax])
```

**Example: Setting axis limits**

```
x = 0:0.1:10;
y = x.^2;
plot(x,y)
axis([0 10 0 100])
grid on
```

**Explanation**

The command axis ([0 10 0 100]) sets x-axis range from 0 to 10, and y-axis range from 0 to 100.

Axis control improves visualization and focuses on specific regions of interest in the graph.

## Summary of MATLAB plotting and visualization

In this section, various plotting and visualization techniques available in MATLAB have been discussed. Different graphical representations including line plots, scatter plots, bar graphs, histograms, pie charts, and three-dimensional plots were explained with suitable examples.

The section also discussed graph customization techniques such as titles, labels, legends, axis control, subplots, and figure exporting methods. MATLAB visualization tools provide efficient and powerful mechanisms for representing numerical and multidimensional data graphically for scientific, engineering, and analytical applications.

## Plotting and visualization in Python

Python provides powerful libraries for graphical representation and data visualization. Among these libraries, Matplotlib is one of the most widely used tools for generating two-dimensional and three-dimensional plots. It provides a module named pyplot that contains functions for generating different types of graphs. Python visualization libraries help programmers represent numerical data graphically for scientific computing, engineering analysis, statistics, machine learning, and data science applications.

Python plotting tools allow the programmer to generate line plots, customize graphs, visualize multidimensional datasets, create statistical charts, and export graphical outputs in different formats.

Before creating plots in Python, the required plotting library must be imported into the program.

## Importing the Matplotlib library

The pyplot module of Matplotlib is generally imported using the alias plt.

**Syntax**

```
import matplotlib.pyplot as plt
```

**Explanation**

In this statement, matplotlib.pyplot provides plotting functions and plt acts as a shorter alias for easier usage in programs.

After importing the library, plotting commands can be executed using the prefix plt.

## Basic 2D plot in Python

The plot() function of Matplotlib is used to generate two-dimensional line graphs between two datasets.

**Syntax**

```
plt.plot(x,y)
```

where x represents the values along the horizontal axis and y represents the values along the vertical axis.

**Example: Plotting a simple line graph**

```
import matplotlib.pyplot as plt
x = [0,1,2,3,4,5]
y = [0,1,4,9,16,25]
plt.plot(x,y)
plt.grid()
plt.show()
```

**Explanation**

In this example, the list x contains values along the horizontal axis, the list y contains the corresponding squared values, plt.plot(x,y) generates the graph, plt.grid() displays grid lines, and plt.show() displays the graphical output, as shown in Figure 8.13. The resulting graph represents the quadratic relationship:

$$y = x^2$$

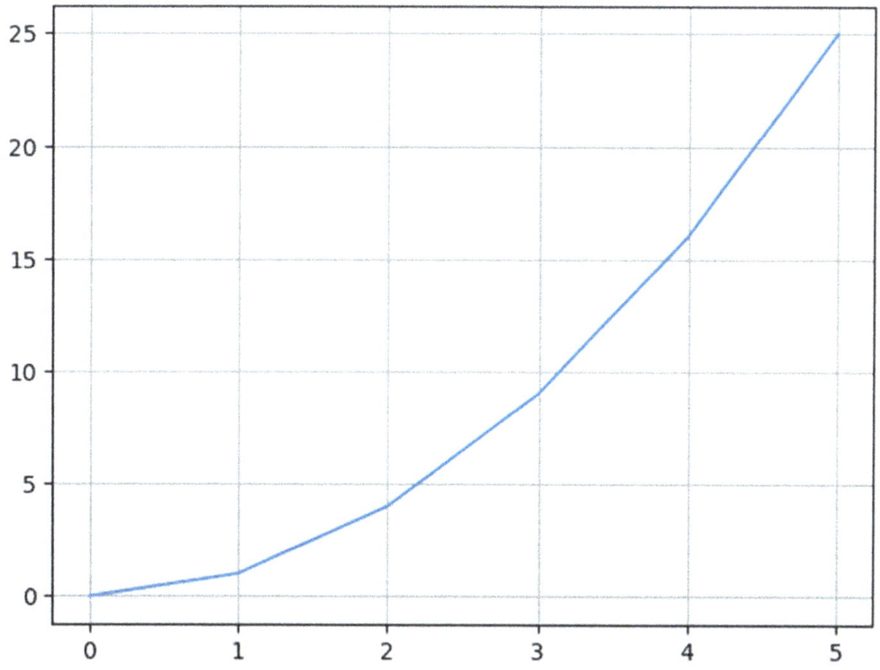


FIGURE 8.13 Plot of parabola on a limited domain in Python.

### Important plotting commands in Python

Some commonly used plotting commands in Matplotlib are presented in Table 8.5.

TABLE 8.5 Commands in Matplotlib for Python.

| Command | Purpose |
|---|---|
| **plt.plot()** | Generates line plots |
| **plt.scatter()** | Creates scatter plots |
| **plt.bar()** | Generates bar graphs |
| **plt.hist()** | Creates histograms |
| **plt.pie()** | Generates pie charts |
| **plt.title()** | Adds graph title |
| **plt.xlabel()** | Labels x-axis |
| **plt.ylabel()** | Labels y-axis |
| **plt.legend()** | Displays legends |
| **plt.grid()** | Displays grid lines |
| **plt.show()** | Displays the graph |

**Advantages of Python visualization libraries**
Python visualization libraries provide flexible and powerful tools for representing numerical data graphically. These libraries support scientific computing, machine learning, statistical analysis, and interactive visualization.

## Customizing plots in Python

Python provides several options for customizing graphical outputs to improve readability and presentation quality. Graphs can be customized by modifying titles, axis labels, line colors, line styles, markers, legends, and grid settings.

Customization helps distinguish multiple datasets and enhances the interpretation of graphical representations.

### Adding titles and axis labels

The Matplotlib library provides functions for adding titles and axis labels to graphs. The common commands used for labeling the axis are provided in Table 8.6.

TABLE 8.6 Common title and axis commands for Python.

| Command | Purpose |
| --- | --- |
| **plt.title()** | Adds graph title |
| **plt.xlabel()** | Labels x-axis |
| **plt.ylabel()** | Labels y-axis |

**Example: Plot with labels and title**

```
import matplotlib.pyplot as plt
x = [0,1,2,3,4,5]
y = [0,1,4,9,16,25]
plt.plot(x,y)
plt.title('Graph of y = x^2')
plt.xlabel('X-axis')
plt.ylabel('Y-axis')
plt.grid()
plt.show()
```

**Explanation**

In this example, plt.title() adds the graph title, plt.xlabel() labels the horizontal axis, plt.ylabel() labels the vertical axis, and plt.grid() displays grid lines. See the outcome in Figure 8.14.

These commands improve the readability and appearance of the graph.

### Modifying line color and style

Python allows customization of line appearance using colors, line styles, markers, and line widths. The list of commonly occurring color codes, line style and markers allowed in Python are tabulated in Tables 8.7–8.9, respectively.

TABLE 8.7 Common color codes in Python.

| Code | Color |
|---|---|
| 'r' | Red |
| 'g' | Green |
| 'b' | Blue |
| 'k' | Black |
| 'm' | Magenta |
| 'c' | Cyan |

TABLE 8.8 Common line styles in Python.

| Style | Representation |
|---|---|
| '-' | Solid line |
| '–' | Dashed line |
| ':' | Dotted line |
| '-.' | Dash-dot line |

TABLE 8.9 Common marker codes in Python.

| Marker | Symbol |
|---|---|
| 'o' | Circle |
| '*' | Star |
| '+' | Plus |
| 'x' | Cross |
| 's' | Square |

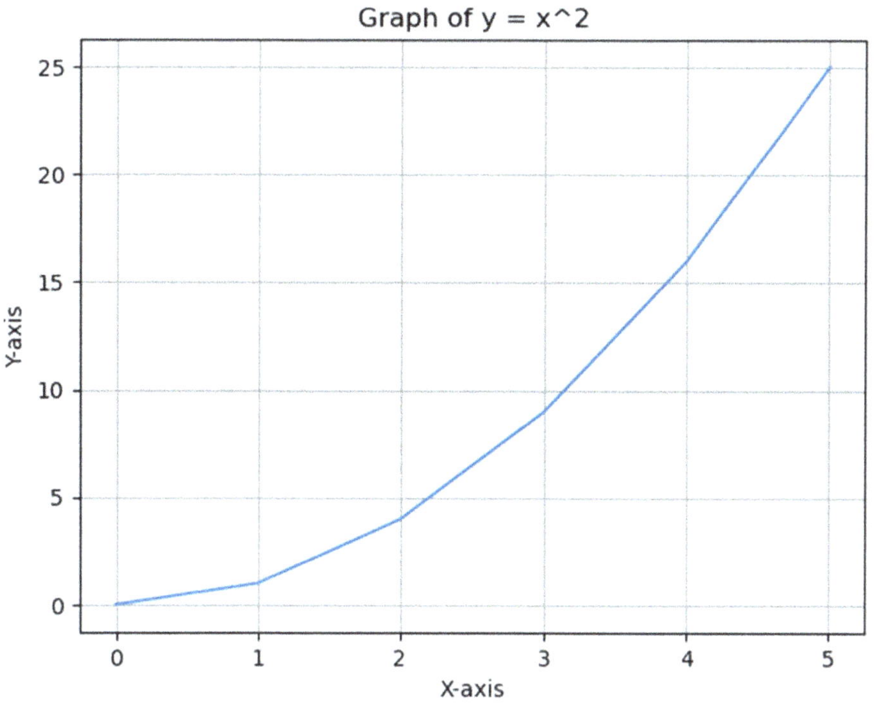


FIGURE 8.14 Plot with the axis labels and titles in Python.

**Example: Customized plot**

```
import matplotlib.pyplot as plt
x = [0,1,2,3,4,5]
y = [0,1,4,9,16,25]
plt.plot(x,y,'r–o',linewidth = 2)
plt.title('Customized Plot')
plt.xlabel('X-axis')
plt.ylabel('Y-axis')
plt.grid()
plt.show()
```

**Explanation**

The outcome is shown in Figure 8.15. In the plotting command 'r–o', r specifies red color, – specifies dashed line style, and o specifies circular markers. The argument linewidth=2, increases the thickness of the plotted curve. The resulting graph appears as a red dashed curve with circular markers.

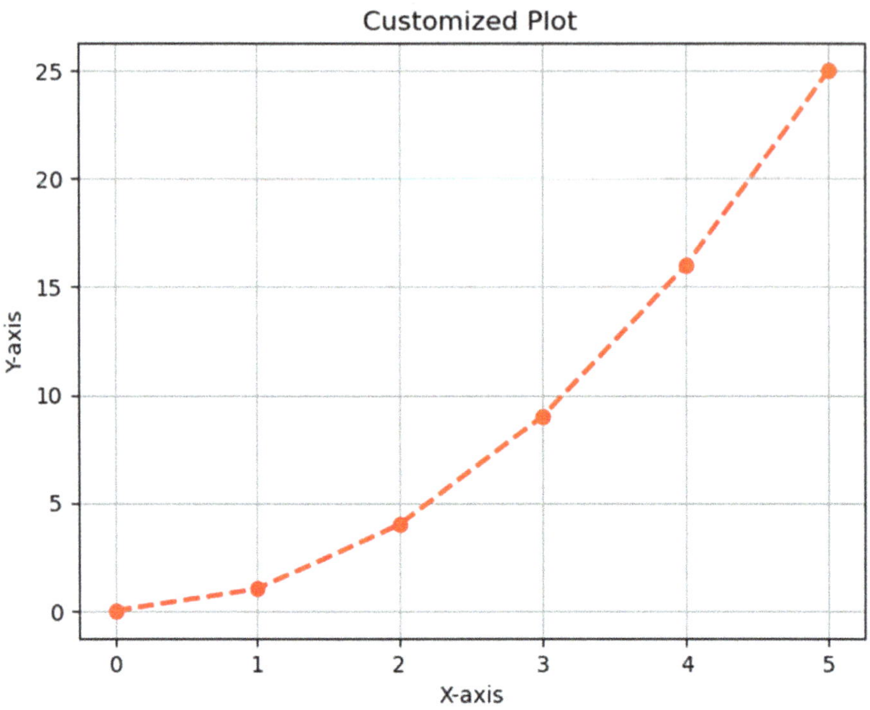


FIGURE 8.15 Enhancement of the graph in Python using markers, line style, and color options.

## Plotting multiple curves in Python

Python allows multiple graphs to be displayed within the same plotting window. This feature is useful for comparing different datasets or mathematical functions.

**Example: Multiple curves on the same graph**

```
import matplotlib.pyplot as plt
x = [-5,-4,-3,-2,-1,0,1,2,3,4,5]
y1 = x*2;
y2 = [0,1,8,27,64,125]
plt.plot(x,y1,label = 'y = x^2')
plt.plot(x,y2,'r–',label = 'y = x^3')
plt.title('Multiple Curves')
plt.xlabel('X-axis')
plt.ylabel('Y-axis')
plt.legend()
```

```
plt.grid()
plt.show()
```

**Explanation**

In this example, the output is shown in Figure 8.16. The first graph represents $y = x^2$, the second graph represents $y = x^3$, plt.legend() identifies the individual curves, different line styles, and colors distinguish the graphs visually.

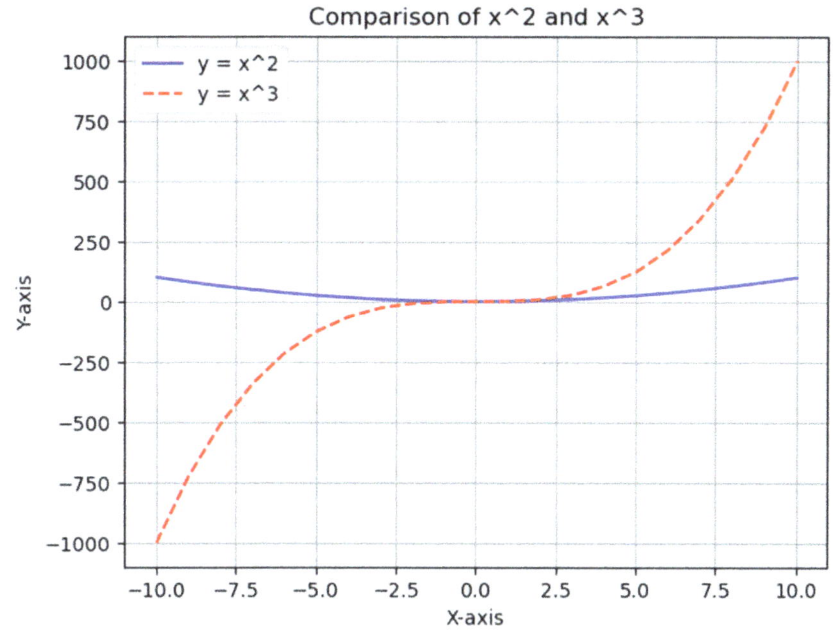


FIGURE 8.16 Plot of two different graphs for the same domain in a single graph.

Multiple plotting improves comparative analysis of datasets and mathematical functions.

## Subplots in Python

In many applications, multiple graphs are required to be displayed within the same figure window for comparison and analysis. Python provides the subplot() function in Matplotlib for dividing the figure into multiple plotting regions. Each subplot acts as an independent plotting area where separate graphs can be displayed.

**Syntax of subplot()**

```
plt.subplot(m,n,p)
```

where m represents the number of rows, n represents the number of columns, and p represents the position of the current plot.

**Example: Creating multiple subplots**

```
import matplotlib.pyplot as plt
import numpy as np
x = np.arange(0,10,0.1)
y1 = np.sin(x)
y2 = np.cos(x)
plt.subplot(2,1,1)
plt.plot(x,y1)
plt.title('Sine Function')
plt.grid()
plt.subplot(2,1,2)
plt.plot(x,y2)
plt.title('Cosine Function')
plt.grid()
plt.show()
```

**Explanation**

See the outcome in Figure 8.17. In the command, plt.subplot(2,1,1) divides the figure into two rows and one column and activates the first plotting region, and plt.subplot(2,1,2) activates the second plotting region. The first subplot displays the sine function, whereas the second subplot displays the cosine function.

## Scatter plots in Python

Scatter plots are used to represent individual data points graphically. They are widely used in statistics, machine learning, and data analysis for studying relationships between variables.

Python provides the scatter() function for generating scatter plots.

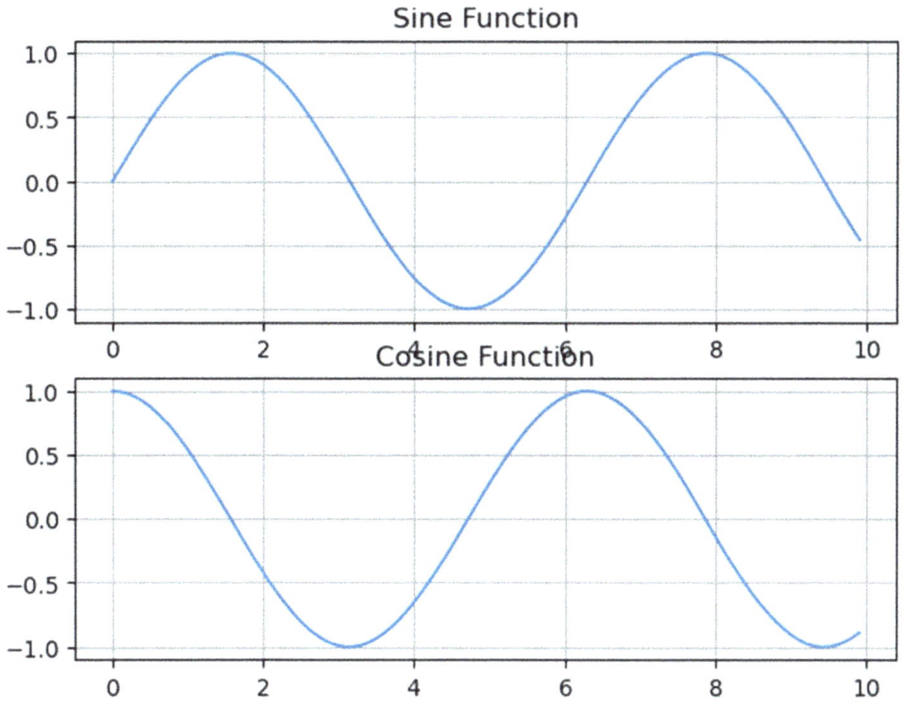


FIGURE 8.17 Plot of two graphs in a single figure window using subplot option in Python.

**Syntax**

```
plt.scatter(x,y)
```

**Example: Scatter plot**

```
import matplotlib.pyplot as plt
x = [1,2,3,4,5,6]
y = [2,4,5,4,6,7]
plt.scatter(x,y)
plt.title('Scatter Plot')
plt.xlabel('X-axis')
plt.ylabel('Y-axis')
plt.grid()
plt.show()
```

**Explanation**

In this example, plt.scatter(x,y) plots the individual data points, each point represents a pair of values from x and y. The graph is presented in Figure 8.18.

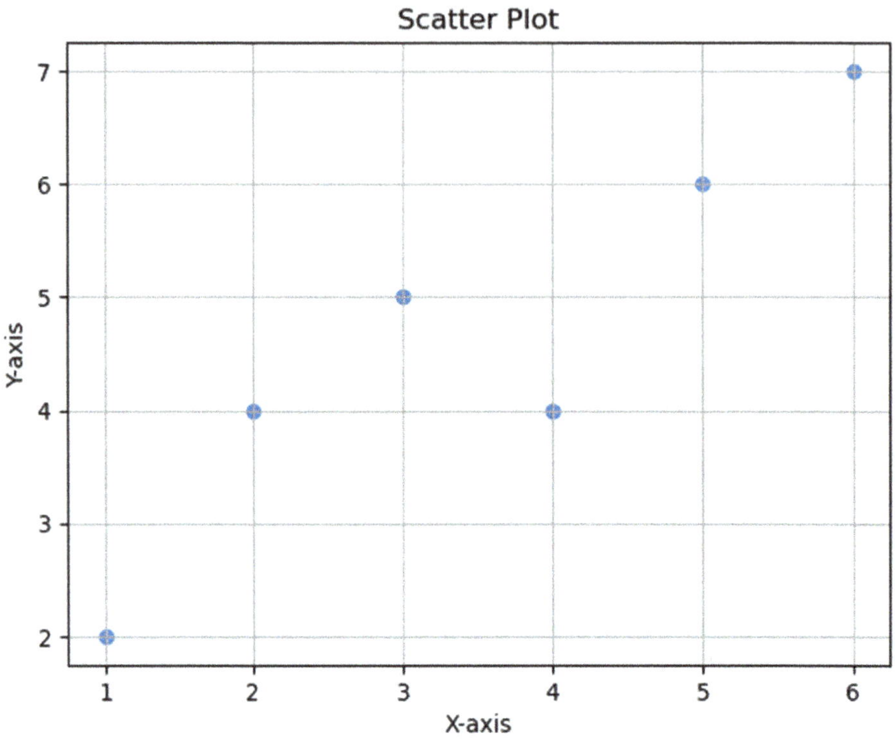


FIGURE 8.18 Scatter plot in Python.

### Bar graphs in Python

Bar graphs are used for representing categorical or discrete data using rectangular bars. The height of each bar corresponds to the magnitude of the data value.

Python provides the bar() function for generating bar graphs.

**Syntax**

```
plt.bar(x,y)
```

**Example: Bar graph**

```
import matplotlib.pyplot as plt
x = [1,2,3,4,5]
y = [5,7,3,8,6]
plt.bar(x,y)
```

```
plt.title('Bar Graph')
plt.xlabel('Category')
plt.ylabel('Values')
plt.grid()
plt.show()
```

**Explanation**

In this example, the values stored in x represent categories, the values stored in y determine the heights of the bars, and plt.bar(x,y) generates the graphical representation. The output of the commands is shown in Figure 8.19.

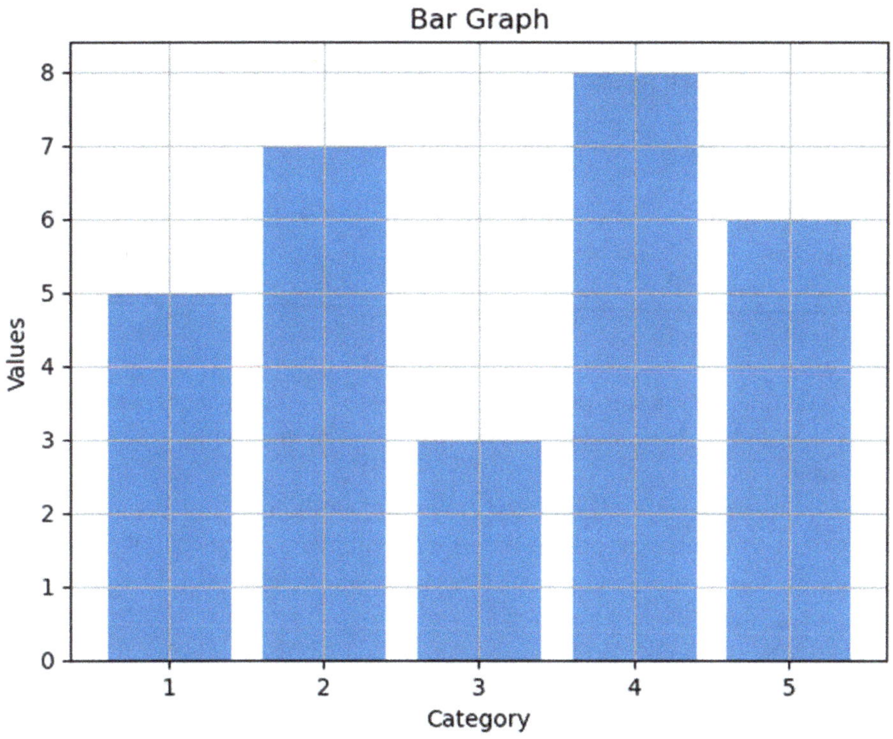


FIGURE 8.19 Bar graph in Python.

### Histograms in Python

Histograms are used for visualizing the frequency distribution of numerical data. They divide the data into intervals called bins and display the frequency of values within each interval.

Python provides the hist() function for generating histograms.

**Syntax**

plt.hist(data)

**Example: Histogram plot**

```
import matplotlib.pyplot as plt
import numpy as np
data = np.random.randn(1,000)
plt.hist(data)
plt.title('Histogram')
plt.xlabel('Values')
plt.ylabel('Frequency')
plt.grid()
plt.show()
```

**Explanation**

In this example, np.random.randn(1,000) generates random values with normal distribution, and plt.hist(data) groups the data into bins and displays the frequency distribution.

Histograms are widely used in statistics and probability analysis for studying data distributions. The output as per the mentioned command is shown in Figure 8.20.

### Pie charts in Python

Pie charts are circular graphical representations used to display the proportional distribution of data. Each sector of the pie represents a fraction or percentage of the total dataset.

Pie charts are widely used in statistics, surveys, business analysis, and comparative data representation.

Python provides the pie() function in Matplotlib for generating pie charts.

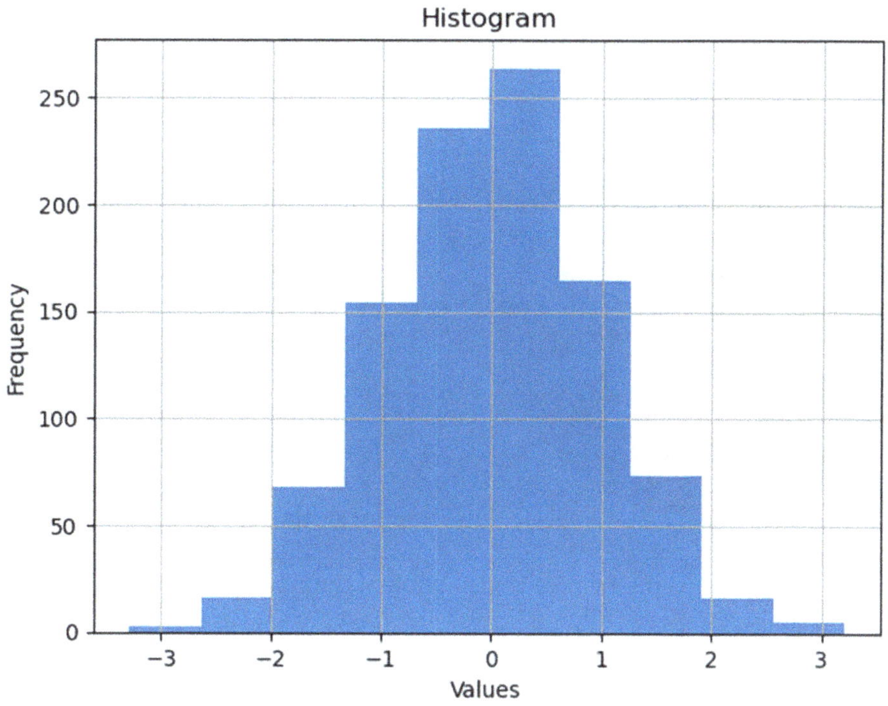


FIGURE 8.20 Plot of histogram in Python.

**Syntax**

```
plt.pie(data)
```

**Example: Pie chart**

```
import matplotlib.pyplot as plt
data = [25,15,30,10,20]
plt.pie(data)
plt.title('Pie Chart')
plt.show()
```

**Explanation**

In this example, the list data contains numerical values and plt.pie(data) divides the circular graph into sectors proportional to the data values. The resultant pie graph is shown in Figure 8.21.

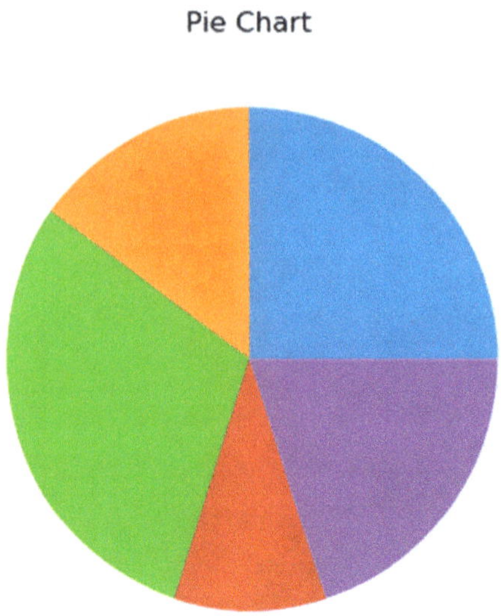


FIGURE 8.21
Pie chart in Python.

Each sector represents the relative contribution of the corresponding value to the complete dataset.

**Adding labels to pie charts**

Labels can be added to identify different sectors of the pie chart.

**Example: Pie chart with labels**

```
import matplotlib.pyplot as plt
data = [20,30,25,25]
labels = ['A','B','C','D']
plt.pie(data,labels = labels)
plt.title('Pie Chart with Labels')
plt.show()
```

Labels improve readability and interpretation of the graphical representation.

### 3D plotting in Python

Python also supports three-dimensional plotting using the mplot3d toolkit available in Matplotlib. Three-dimensional visualization is useful for scientific simulations, engineering analysis, mathematical modeling, and multidimensional data representation.

Commonly used 3D plotting functions in Python include plot3D(), plot_surface() and scatter3D().

**Importing the 3D plotting toolkit**
Before generating 3D plots, the required toolkit must be imported.

**Syntax**
```
from mpl_toolkits.mplot3d import Axes3D
```

**3D line plot in Python**
A three-dimensional line graph can be generated using the plot3D() function.

**Example: 3D line plot**
```
import matplotlib.pyplot as plt
import numpy as np
from mpl_toolkits.mplot3d import Axes3D
fig = plt.figure()
ax = fig.add_subplot(111,projection = '3d')
t = np.arange(0,10,0.1)
x = np.sin(t)
y = np.cos(t)
z = t
ax.plot3D(x,y,z)
ax.set_title('3D Line Plot')
ax.set_xlabel('X-axis')
ax.set_ylabel('Y-axis')
ax.set_zlabel('Z-axis')
plt.show()
```

**Explanation**
In this example, projection='3d' activates the three-dimensional plotting environment, x = sin(t) and y = cos(t) generate circular coordinates, z = t provides height variation, and plot3D() generates the three-dimensional spiral curve. The output is shown in Figure 8.22.

### Surface plots in Python

Surface plots are used for visualizing functions of two variables in three-dimensional form.

Python provides the plot_surface() function for generating surface plots.

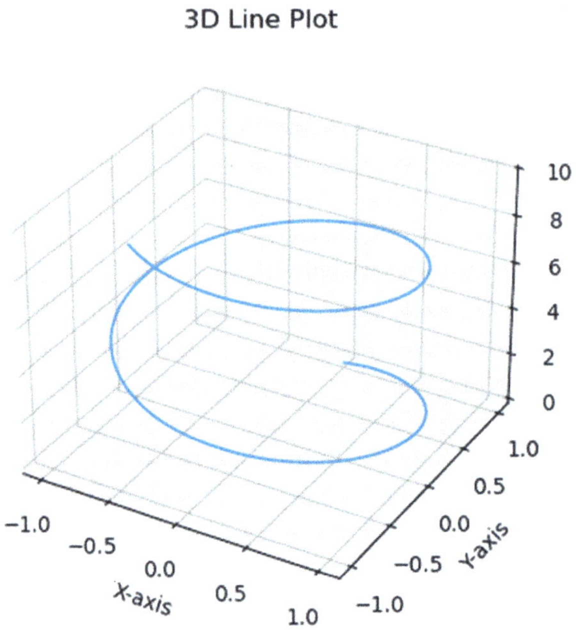


FIGURE 8.22 Three-dimensional spiral curve in Python.

**Example: Surface plot**

```
import matplotlib.pyplot as plt
import numpy as np
from mpl_toolkits.mplot3d import Axes3D
fig = plt.figure()
ax = fig.add_subplot(111,projection = '3d')
x = np.arange(-5,5,0.5)
y = np.arange(-5,5,0.5)
X,Y = np.meshgrid(x,y)
Z=X**2+Y**2
ax.plot_surface(X,Y,Z)
ax.set_title('Surface Plot')
ax.set_xlabel('X-axis')
ax.set_ylabel('Y-axis')
ax.set_zlabel('Z-axis')
plt.show()
```

**Explanation**

In this example, meshgrid() generates coordinate matrices, Z = X**2 + Y**2 computes the surface values, and plot_surface() generates the three-dimensional surface plot.

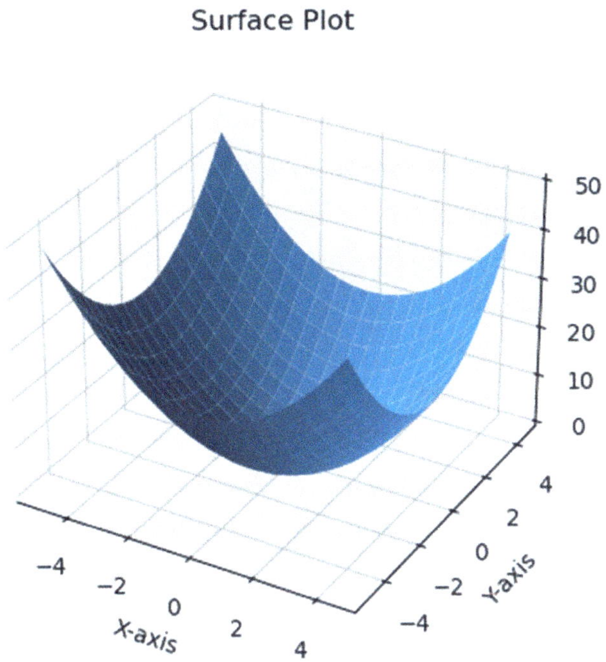


FIGURE 8.23 Surf plot in Python.

The resulting graph represents a paraboloid surface, as shown in Figure 8.23.

## Saving and exporting plots in Python

Python allows graphical outputs to be saved in different file formats for reports, presentations, research publications, and documentation purposes. The generated plots can be stored as image files such as: PNG, JPG, PDF, and SVG formats.

Saving graphical outputs helps preserve visualization results for future use and analysis.

**Saving plots using savefig()**

The savefig() function of Matplotlib is used to save the current figure in different formats.

**Syntax**

```
plt.savefig('filename.extension')
```

where filename specifies the name of the file, and extension specifies the output file format.

**Example: Saving a plot as PNG image**

```
import matplotlib.pyplot as plt
x = [0,1,2,3,4,5]
y = [0,1,4,9,16,25]
plt.plot(x,y)
plt.title('Quadratic Curve')
plt.savefig('quadratic_plot.png')
plt.show()
```

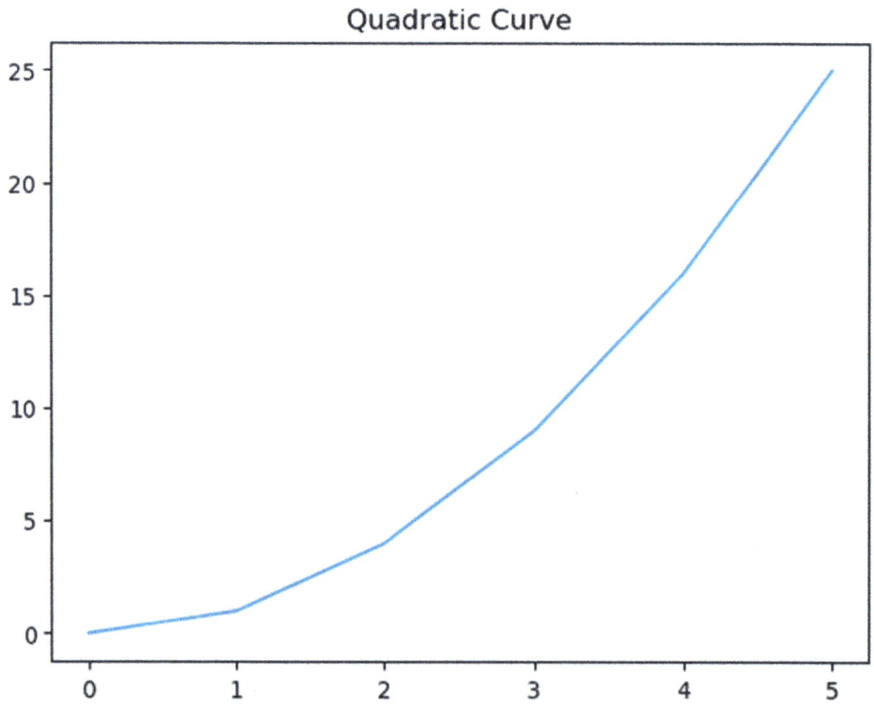


FIGURE 8.24 Saving a graph in .png format in Python.

**Explanation**
In this example, plt.plot(x,y) generates the graph, and plt.savefig('quadratic_plot.png') stores the graphical output as a PNG image file in the system with the name provided. The output is shown in Figure 8.24.

### Figure customization in Python

Python provides several commands for customizing figure windows and improving graphical presentation. Some commonly used customization commands are provided in Table 8.10.

TABLE 8.10 Commonly used commands of plots in Python.

| Command | Purpose |
| --- | --- |
| **plt.figure()** | Creates a new figure window |
| **plt.legend()** | Displays graph legends |
| **plt.grid()** | Displays grid lines |
| **plt.xlim()** | Sets x-axis limits |
| **plt.ylim()** | Sets y-axis limits |

**Example: Figure and axis customization**
```
import matplotlib.pyplot as plt
x = [0,1,2,3,4,5]
y1 = [0,1,4,9,16,25]
y2 = [0,1,8,27,64,125]
plt.figure(figsize = (6,4))
plt.plot(x,y1,label = 'y = x^2')
plt.plot(x,y2,'r–',label = 'y = x^3')
plt.title('Customized Figure')
plt.xlabel('X-axis')
plt.ylabel('Y-axis')
plt.xlim(0,5)
plt.ylim(0,130)
plt.legend()
plt.grid()
plt.show()
```

**Explanation**

In this example, figure(figsize = (6,4)) specifies the figure dimensions, xlim() controls the horizontal axis range, ylim() controls the vertical axis range, legend() identifies different curves, and grid() improves readability of the graph.

These customization commands enhance the graphical appearance and analytical interpretation of plots, as shown in Figure 8.25.

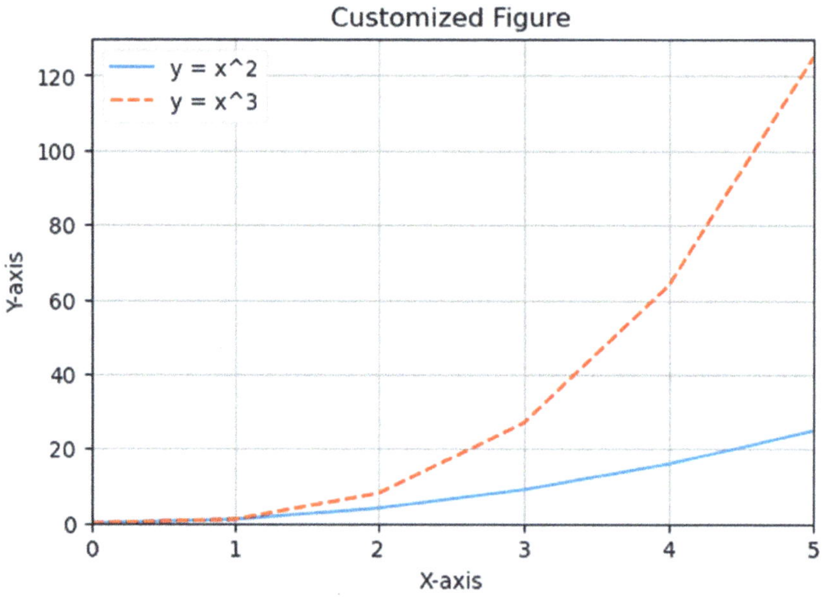


FIGURE 8.25 Creation of graph with customization of axis and figure in Python.

## Summary of Python plotting and visualization

In this section, various plotting and visualization techniques available in Python were discussed using the Matplotlib library. Different graphical representations including line plots, scatter plots, bar graphs, histograms, pie charts, and three-dimensional plots were explained with suitable examples.

This section also discussed graph customization techniques such as titles, labels, legends, subplots, axis control, and figure exporting methods. Python visualization libraries provide efficient and flexible mechanisms for representing numerical and multidimensional data graphically for scientific, engineering, and analytical applications.

## Overall summary

This chapter presents the plotting and visualization functionalities of MATLAB and Python. We generated various two-dimensional and three-dimensional graphical representations with built-in functions and visualization frameworks. In this chapter, we illustrated the application of graphical tools to aid programmers in data analysis, the interpretation of mathematical relationships, and the effective presentation of computing outcomes. These visualization approaches are fundamental to scientific computing, engineering analysis, and contemporary data-driven applications.

## Exercise

### Section A: MATLAB plotting and visualization

1. Write a MATLAB program to plot the graph of:
   $y = x^2$
   for $x = 0$ to 10.
2. Create a MATLAB program to plot sine and cosine curves on the same graph.
3. Write a MATLAB program to generate a scatter plot for two datasets.
4. Create a bar graph representing the marks of five students.
5. Write a MATLAB program to generate a histogram using random data.
6. Create a pie chart showing the percentage distribution of four categories.
7. Write a MATLAB program using subplot() to display:
   – sine function,
   – cosine function,
   in separate plotting regions.
8. Generate a three-dimensional spiral curve using plot3().
9. Create a surface plot for the function:
   $$z = x^2 + y^2$$
10. Write a MATLAB program to save a generated plot as a PNG image file.

### Section B: Python plotting and visualization

1. Write a Python program to plot the graph of:
   $$y = x^2$$
2. Create a Python program to plot sine and cosine curves on the same graph.
3. Write a Python program to generate a scatter plot for two datasets.
4. Create a bar graph representing monthly sales data.
5. Write a Python program to generate a histogram using random numbers.

6. Create a pie chart showing percentage distribution of students in different courses.
7. Write a Python program using subplot() to display two different mathematical functions.
8. Generate a three-dimensional spiral curve using plot3D().
9. Create a surface plot for the function:

$$z = x^2 + y^2$$

10. Write a Python program to save a generated plot as a PNG image file.

## Computational practice questions with short answers

**Q1. Plot $y = x^2$ for $-5 \leq x \leq 5$.**

**MATLAB**

```
x = -5:0.1:5;
y = x.^2;
plot(x,y)
grid on
title('Plot of y = x^2')
```

**Python**

```
import numpy as np
import matplotlib.pyplot as plt

x = np.linspace(-5,5,100)
y = x**2
plt.plot(x,y)
plt.grid(True)
plt.title("Plot of y = x^2")
plt.show()
```

**Q2. Plot $y = sinx$ for $0 \leq x \leq 2\pi$.**

**MATLAB**

```
x = 0:0.01:2*pi;
y = sin(x);
plot(x,y)
grid on
```

### Python

```
x = np.linspace(0,2*np.pi,200)
y = np.sin(x)
plt.plot(x,y)
plt.grid(True)
plt.show()
```

**Q3. Plot $y = cosx$ for $0 \le x \le 2\pi$.**

### MATLAB

```
x = 0:0.01:2*pi;
plot(x,cos(x))
grid on
```

### Python

```
x = np.linspace(0,2*np.pi,200)
plt.plot(x,np.cos(x))
plt.grid(True)
plt.show()
```

Q4. Plot $y = sinx$ and $y = cosx$ together.

### MATLAB

```
x = 0:0.01:2*pi;
plot(x,sin(x),x,cos(x))
legend('sin(x)','cos(x)')
grid on
```

### Python

```
x = np.linspace(0,2*np.pi,200)
plt.plot(x,np.sin(x),label = 'sin(x)')
plt.plot(x,np.cos(x),label = 'cos(x)')
plt.legend()
plt.grid(True)
plt.show()
```

## Q5. Plot $y = e^x$ for $0 \le x \le 5$.

### MATLAB

```
x = 0:0.1:5;
y = exp(x);
plot(x,y)
grid on
```

### Python

```
x = np.linspace(0,5,100)
y = np.exp(x)
plt.plot(x,y)
plt.grid(True)
plt.show()
```

## Q6. Plot $y = e^{-x}$ for $0 \le x \le 5$.

### MATLAB

```
x = 0:0.1:5;
plot(x,exp(-x))
grid on
```

### Python

```
x = np.linspace(0,5,100)
plt.plot(x,np.exp(-x))
plt.grid(True)
plt.show()
```

## Q7. Plot $y = logx$ for $0.1 \le x \le 10$.

### MATLAB

```
x = 0.1:0.1:10;
plot(x,log(x))
grid on
```

### Python

```
x = np.linspace(0.1,10,100)
plt.plot(x,np.log(x))
plt.grid(True)
plt.show()
```

### Q8. Plot $y = |x|$ for $-5 \leq x \leq 5$.

### MATLAB

```
x = -5:0.1:5;
plot(x,abs(x))
grid on
```

### Python

```
x = np.linspace(-5,5,100)
plt.plot(x,np.abs(x))
plt.grid(True)
plt.show()
```

Q9. Plot the cubic function $y = x^3$.

### MATLAB

```
x = -5:0.1:5;
plot(x,x.^3)
grid on
```

### Python

```
x = np.linspace(-5,5,100)
plt.plot(x,x**3)
plt.grid(True)
plt.show()
```

## Q10. Plot $y = x^2 - 4x + 3$.

### MATLAB

```
x = -2:0.1:6;
y = x.^2 - 4*x + 3;
plot(x,y)
grid on
```

### Python

```
x = np.linspace(-2,6,100)
y = x**2 - 4*x + 3
plt.plot(x,y)
plt.grid(True)
plt.show()
```

Q11. Plot the parametric circle $x = \cos t, y = \sin t$.

### MATLAB

```
t = 0:0.01:2*pi;
x = cos(t);
y = sin(t);
plot(x,y)
axis equal
grid on
```

### Python

```
t = np.linspace(0,2*np.pi,200)
x = np.cos(t)
y = np.sin(t)
plt.plot(x,y)
plt.axis('equal')
plt.grid(True)
plt.show()
```

Q12. Plot the parametric spiral $x = tcost, y = tsint$.

**MATLAB**

```
t = 0:0.1:10*pi;
x = t.*cos(t);
y = t.*sin(t);
plot(x,y)
grid on
```

**Python**

```
t = np.linspace(0,10*np.pi,500)
x = t*np.cos(t)
y = t*np.sin(t)
plt.plot(x,y)
plt.grid(True)
plt.show()
```

Q13. Plot the numerical derivative of $y = sinx$.

**MATLAB**

```
x = 0:0.01:2*pi;
y = sin(x);
dy = gradient(y,x);
plot(x,y,x,dy)
legend('sin(x)','Derivative')
grid on
```

**Python**

```
x = np.linspace(0,2*np.pi,200)
y = np.sin(x)
dy = np.gradient(y,x)
plt.plot(x,y,label = 'sin(x)')
plt.plot(x,dy,label = 'Derivative')
plt.legend()
plt.grid(True)
plt.show()
```

Q14. Plot the numerical integration area under $y = x^2$.

**MATLAB**

```
x = 0:0.1:3;
y = x.^2;
area(x,y)
grid on
```

**Python**

```
x = np.linspace(0,3,100)
y = x**2
plt.fill_between(x,y)
plt.grid(True)
plt.show()
```

**Q15. Plot the histogram of random normal data.**

**MATLAB**

```
data = randn(1,000,1);
histogram(data)
grid on
```

**Python**

```
data = np.random.randn(1,000)
plt.hist(data)
plt.grid(True)
plt.show()
```

Q16. Create a scatter plot for $x$ and $y = x^2$ + noise.

**MATLAB**

```
x = 1:50;
y = x.^2 + 100*randn(size(x));
scatter(x,y)
grid on
```

### Python

```
x = np.arange(1,51)
y = x**2 + 100*np.random.randn(len(x))
plt.scatter(x,y)
plt.grid(True)
plt.show()
```

## Q17. Plot the bar graph of values [4,7,2,9,5].

### MATLAB

```
y = [4 7 2 9 5];
bar(y)
grid on
```

### Python

```
y = [4,7,2,9,5]
plt.bar(range(1,6),y)
plt.grid(True)
plt.show()
```

Q18. Plot the stem graph of sequence $a_n = n^2$, $n = 1, \ldots, 10$.

### MATLAB

```
n = 1:10;
a = n.^2;
stem(n,a)
grid on
```

### Python

```
n = np.arange(1,11)
a = n**2
plt.stem(n,a)
plt.grid(True)
plt.show()
```

## Q19. Plot the Fibonacci sequence.

### MATLAB

```
f = [1 1 2 3 5 8 13 21 34 55];
stem(f)
grid on
```

### Python

```
f = [1,1,2,3,5,8,13,21,34,55]
plt.stem(f)
plt.grid(True)
plt.show()
```

## Q20. Plot a 3D surface $z = x^2 + y^2$.

### MATLAB

```
[x,y] = meshgrid(-5:0.2:5);
z = x.^2 + y.^2;
surf(x,y,z)
```

### Python

```
from mpl_toolkits.mplot3d import Axes3D
x = np.linspace(-5,5,50)
y = np.linspace(-5,5,50)
X,Y = np.meshgrid(x,y)
Z = X**2 + Y**2
fig = plt.figure()
ax = fig.add_subplot(111, projection = '3d')
ax.plot_surface(X,Y,Z)
plt.show()
```

Q21. Create a contour plot of $z = x^2 + y^2$.

**MATLAB**

```
[x,y] = meshgrid(-5:0.2:5);
z = x.^2 + y.^2;
contour(x,y,z)
grid on
```

**Python**

```
x = np.linspace(-5,5,50)
y = np.linspace(-5,5,50)
X,Y = np.meshgrid(x,y)
Z = X**2 + Y**2
plt.contour(X,Y,Z)
plt.grid(True)
plt.show()
```

Q22. Plot the vector field $u = -y, v = x$.

**MATLAB**

```
[x,y] = meshgrid(-3:1:3);
u = -y;
v = x;
quiver(x,y,u,v)
axis equal
grid on
```

**Python**

```
x,y = np.meshgrid(np.arange(-3,4),np.arange
(-3,4))
u = -y
v = x
plt.quiver(x,y,u,v)
plt.axis('equal')
plt.grid(True)
plt.show()
```

Q23. Plot the solution of $y' = y, y(0) = 1$, exact solution $y = e^x$.

**MATLAB**

```
x = 0:0.1:2;
y = exp(x);
plot(x,y)
grid on
title('Solution of dy/dx = y')
```

**Python**

```
x = np.linspace(0,2,100)
y = np.exp(x)
plt.plot(x,y)
plt.grid(True)
plt.title("Solution of dy/dx = y")
plt.show()
```

**Q24. Plot the error between exact and approximate solutions.**

**MATLAB**

```
x = 0:0.1:2;
exact = exp(x);
approx = 1 + x + x.^2/2;
error = abs(exact - approx);
plot(x,error)
grid on
```

**Python**

```
x = np.linspace(0,2,100)
exact = np.exp(x)
approx = 1 + x + x**2/2
error = np.abs(exact - approx)
plt.plot(x,error)
plt.grid(True)
plt.show()
```

### Q25. Plot the heatmap of a matrix.

**MATLAB**

```
A = rand(5,5);
imagesc(A)
colorbar
```

**Python**

```
A = np.random.rand(5,5)
plt.imshow(A)
plt.colorbar()
plt.show()
```

## Mini project: mathematical visualization and comparative analysis of numerical solutions of differential equations using MATLAB and Python

### Project title

Visualization and Comparative Analysis of Numerical and Exact Solutions of First-Order Differential Equations Using MATLAB and Python

### Objective

The objective of this project is to numerically solve a first-order differential equation, compare the numerical and exact solutions, visualize errors, and analyze computational results using MATLAB and Python.

### Problem statement

Consider the differential equation:

$$\frac{dy}{dx} = y - x^2 + 1, y(0) = 0.5$$

Solve the problem numerically using Euler's method and compare it with the exact solution.

Exact solution:

$$y(x) = ( + 1)^2 - 0.5e^x$$

### Tasks

---

**Task 1: Generate numerical solution**

- Implement Euler's method
- Compute solution values between $x = 0$ and $x = 2$
- Use step size $h = 0.1$

---

---

**Task 2: Plot numerical versus exact solutions**

Create:

- Exact solution curve
- Numerical solution curve
- Combined comparison graph

---

---

**Task 3: Error visualization**

Calculate:

$$\text{Error} = |y_{\text{exact}} - y_{\text{numerical}}|$$

Plot:

- Absolute error curve
- Error trend with increasing $x$

---

---

**Task 4: Statistical analysis**

Compute:

- Maximum error
- Mean error
- Root mean square error (RMSE)

---

**Task 5: MATLAB-Python comparison**

Compare:

- Computational time
- Plot appearance
- Ease of implementation
- Numerical accuracy

### Expected visualizations

Students should create:

1. Numerical solution plot
2. Exact solution plot
3. Combined comparison plot
4. Error plot
5. Bar graph of error metrics

### Extension activities

- Implement smaller step sizes $h = 0.05, 0.01$
- Compare with higher-order methods
- Produce 3D visualization of error surfaces

CHAPTER 9

# Computational mathematics and data analysis using MATLAB and Python

Computational mathematics and data analysis are fundamental to contemporary scientific computers, engineering applications, and data-informed decision making. Solving the real-world problems across several fields requires precise interpretation of numerical data and effective execution of mathematical calculations. Programming environments like MATLAB and Python offer tools for numerical computing, data processing, statistical analysis, and visualization, enabling users to convert raw data into valuable information.

Numerical computation entails executing arithmetic operations, assessing mathematical expressions, resolving equations, and manipulating numerical datasets. Data analysis focuses on extracting valuable insights from data through statistical measures, summarization techniques, and computing methodologies. Collectively, these competencies constitute the foundation of scientific and analytical computing.

This chapter presents the computational and data analysis capabilities available in MATLAB and Python. This chapter examines several approaches for doing mathematical computations, manipulating numbers, and employing statistical functions to effectively evaluate datasets. Readers can expect acquiring the required skills to execute computational tasks and do fundamental data analysis in both programming environments, utilizing real examples and comparative discussions.

## Working with MATLAB

MATLAB provides a computational environment designed explicitly for numerical computation, mathematical modeling, matrix manipulation, and data analysis. It possesses an extensive library of built-in functions that enable programmers to execute complex mathematical calculations quickly with minimal programming effort. One of MATLAB's significant advantages is its capacity to directly manipulate vectors, matrices, and higher-dimensional data structures. This capability enables users to address substantial computational and data processing challenges with ease and efficiency. MATLAB is widely utilized in engineering, scientific research, machine learning, signal processing, finance, and several analytical applications where computational efficiency and precision are paramount.

 | HTTPS://DOI.ORG/10.1515/9783112207758-009

The computational tools in MATLAB are categorized into two primary domains as numerical computation and data analysis. While the numerical computation refers to the execution of mathematical computations and the resolution of numerical issues, the data analysis entails the extraction of valuable insights from datasets through statistical and analytical techniques.

## Numerical computation in MATLAB

Numerical computation constitutes the foundation of scientific and technical programming. It includes executing arithmetic operations, assessing mathematical statements, resolving equations, and handling numerical data. MATLAB offers an extensive array of operators and built-in functions for mathematical computations, enabling precise and efficient calculations.

In contrast to conventional programming languages that often execute computations on an element-by-element basis, MATLAB is designed for matrix and vector arithmetic. This enables the execution of extensive computing operations using concise and comprehensible commands. Numerical computing in MATLAB encompasses fundamental arithmetic operations as well as intricate mathematical and engineering calculations.

### Mathematical computations using MATLAB

MATLAB contains various inbuilt mathematical functions for the efficient computation of numerical expressions. These routines assist the programmer from the necessity of manually executing standard mathematical operations and ensure the precise computational outcomes. MATLAB is extensively employed for mathematical computations in engineering analysis, scientific modeling, optimization, finance, and data processing applications.

MATLAB encompasses various categories of mathematical functions, including trigonometric, logarithmic, exponential, and rounding functions, as well as arithmetic of complex numbers and specialized mathematical operations. These routines enable the user to do intricate numerical computations with a straightforward and succinct instruction. The brief details of the common mathematical functions are presented in Table 9.1.

TABLE 9.1 Common mathematical functions of MATLAB.

| Functions | Description |
|---|---|
| **sqrt(x)** | Square root |
| **abs(x)** | Absolute value |
| **exp(x)** | Exponential function |
| **log(x)** | Natural logarithm |
| **log10(x)** | Base-10 logarithm |
| **sin(x)** | Sine function |
| **cos(x)** | Cosine function |
| **tan(x)** | Tangent function |
| **round(x)** | Round to nearest integer |
| **floor(x)** | Round towards smaller integer |
| **ceil(x)** | Round towards larger integer |

**Example: Evaluating mathematical functions**

```
x = 16;
a = sqrt(x)
b = abs(-25)
c = exp(2)
d = log(5)
e = round(4.67)
```

**Output**

```
a =
   4
b =
   25
c =
   7.3891
d =
   1.6094
e =
   5
```

**Explanation:**

The function sqrt() calculates the square root, abs() yields the absolute value of a number, exp() determines the exponential function, log() computes the natural logarithm, and round() approximates a number to the nearest integer. These functions are commonly employed in scientific and engineering calculations.

### Trigonometric and exponential computations

Trigonometric and exponential functions are integral to scientific computing, signal processing, numerical analysis, and mathematical modeling. MATLAB offers inherent support for these calculations through highly optimized numerical algorithms.

**Example:**

```
x = pi/4;
s = sin(x)
c = cos(x)
t = tan(x)
e = exp(1)
```

### Complex number computations in MATLAB

Complex numbers commonly occur in various field involving engineering, physics, signal processing, control systems, and electrical circuit analysis. MATLAB offers inherent support for complex arithmetic, enabling direct computations with imaginary numbers without the need for specialized techniques.

A complex number is generally represented as:

$$z = a + bi$$

where $a$ is the real part, $b$ is the imaginary part, with $i = \sqrt{-1}$represents the imaginary unit.

MATLAB recognizes both i and j as the imaginary unit.

### Defining complex numbers

A complex number can be defined directly using the imaginary unit.

**Example:**

```
z1 = 3 + 4i
z2 = 2 – 5i
```

**Output:**

```
z1 =
   3.0000 + 4.0000i
z2 =
   2.0000 – 5.0000i
```

### Basic operations on complex numbers

MATLAB supports arithmetic operations on complex numbers in the same manner as real numbers.

**Example:**

```
z1 = 3 + 4i;
z2 = 2 – 5i;
sum_z = z1 + z2
prod_z = z1 * z2
```

**Output:**

```
sum_z =
   5.0000 – 1.0000i
prod_z =
   26.0000 – 7.0000i
```

### Extracting real and imaginary parts

MATLAB provides dedicated functions for obtaining the real and imaginary components of a complex number. A list of important functions frequently used with complex numbers is detailed in Table 9.2.

TABLE 9.2 Mathematical functions for the complex numbers in MATLAB.

| Functions | Purpose |
|---|---|
| **real(z)** | Real part |
| **imag(z)** | Imaginary part |
| **conj(z)** | Complex conjugate |
| **abs(z)** | Magnitude |
| **angle(z)** | Phase angle |

**Example:**

```
z = 3 + 4i;
real_part = real(z)
imag_part = imag(z)
magnitude = abs(z)
conjugate = conj(z)
```

**Output:**

```
real_part =
  3
imag_part =
  4
magnitude =
  5
conjugate =
  3.0000 - 4.0000i
```

**Explanation:**
The function real() extracts the real component, imag() extracts the imaginary component, abs() computes the magnitude of the complex number, and conj() computes the complex conjugate.

Complex number computations are extensively used in electrical engineering, communication systems, signal analysis, and scientific simulations.

## Polynomial computations in MATLAB

Polynomials commonly occur in mathematics, engineering, physics, control systems, and numerical analysis. MATLAB provides various inbuilt functions for defining, evaluating, manipulating, and determining the roots of polynomial expressions. These functions streamline polynomial computations and obviate the necessity for protracted human calculations.

A polynomial of degree $n$ can be represented as:

$$P(x) = a_n x^n + a_{n-1} x^{n-1} + \cdots + a_1 x + a_0$$

In MATLAB, a polynomial is represented by a vector containing its coefficients arranged in descending powers of the variable.

### Defining a polynomial

Consider the polynomial:

$$P(x) = x^3 - 6x^2 + 11x - 6$$

The coefficient vector is defined as:

P = [1 -6 11 -6];

The coefficients correspond to the powers:

$$1x^3 - 6x^2 + 11x - 6$$

### Evaluating a polynomial

MATLAB provides the polyval() function to evaluate a polynomial at a specified value of x.

**Syntax:**

polyval(P,x)

**Example:**

```
P = [1 -6 11 -6];

value = polyval(P,2)
```

**Output:**

```
value =

    0
```

**Explanation:**

The command evaluates $P(2) = 2^3 - 6(2)^2 + 11(2) - 6$, which results in zero.

### Finding polynomial roots

One of the most useful polynomial operations is determining the roots of a polynomial. MATLAB provides the roots() function for this purpose.

**Example:**

```
P = [1 -6 11 -6];

r = roots(P)>
```

**Output:**

```
r =

    3
    2
    1
```

**Explanation:**
The roots represent the values of x for which $P(x) = 0$
Thus, the polynomial has roots at $x = 1$, $x = 2$, and $x = 3$.

### Constructing a polynomial from roots

MATLAB can also construct a polynomial when the roots are known.

**Example:**

```
r = [1 2 3];

P = poly(r)
```

**Output:**

```
P =

    1   -6   11   -6
```

**Explanation**
The function poly() generates the coefficient vector corresponding to the polynomial having the specified roots.

## Numerical integration and differentiation in MATLAB

Many scientific and engineering problems require the evaluation of derivatives and integrals. MATLAB provides built-in numerical methods that allow differentiation and integration of data and functions efficiently.

Numerical differentiation is used to determine rates of change, whereas numerical integration is used to estimate areas, accumulated quantities, and total effects.

### Numerical differentiation

The diff() function computes the difference between consecutive elements of a vector and can be used to approximate derivatives.

**Example:**

```
x = [1 2 4 7 11];
d = diff(x)
```

**Output:**

```
d =
   1   2   3   4
```

**Explanation**

The output represents which gives the differences between consecutive elements.

### Numerical integration using trapezoidal rule

MATLAB provides the trapz() function for numerical integration based on the trapezoidal rule.

**Example:**

```
x = 0:0.5:2;

y = x.^2;

I = trapz(x,y)
```

**Output**

```
I =

   2.7500
```

**Explanation**

The command trapz(x,y) approximates the area under the curve $y = x^2$ over the interval [0, 2]. Numerical integration is particularly useful when analytical integration is difficult or when only discrete data points are available.

## Data analysis in MATLAB

Data analysis involves the analyzing, cleansing, transformation, and modeling of data to discover valuable insights, draw conclusions, and facilitate decision-making. In scientific and engineering computing applications, data analysis is employed to identify patterns, trends, variances, and relationships within datasets. MATLAB provides statistical tools to enhance the analysis of numerical data. The initial phase in dataset analysis involves the calculation of descriptive statistics. MATLAB includes many inbuilt functions that compute mean, median, maximum and minimum values, variance, standard deviation, and other related statistical functions.

### Descriptive statistical measures

Descriptive statistics summarize a dataset using numerical measures that describe its central tendency and distribution.

Consider the following dataset:

X = [12 15 18 21 24 18 15 30];

#### Mean of a dataset

The mean represents the arithmetic average of all observations in a dataset.

**Syntax**

```
mean(X)
```

**Example:**

X = [12 15 18 21 24 18 15 30];

M = mean(X)

**Output:**

M =

19.125

**Explanation**

The mean is computed as:

$$\text{Mean} = \frac{\text{Sum of observations}}{\text{Number of observations}}$$

The mean provides a measure of the central location of the dataset.

### Median of a dataset

The median represents the middle value when the observations are arranged in ascending order.

**Syntax**

```
median(X)
```

**Example:**

X = [12 15 18 21 24 18 15 30];

Md = median(X)

**Output:**

Md =

18

**Explanation:**

The median is less sensitive to extreme values and therefore provides a robust measure of central tendency.

### Mode of a dataset

The mode represents the value that occurs most frequently in the dataset.

**Syntax**

```
mode(X)
```

**Example:**

```
X = [12 15 18 21 24 18 15 30];

Mo = mode(X)
```

**Output**

```
Mo =

   15
```

**Explanation**

The mode identifies the most frequently occurring observation in the dataset.

### Maximum and minimum values

The largest and smallest values in a dataset often provide important information regarding the range and spread of the data.

**Example:**

```
X = [12 15 18 21 24 18 15 30];

maximum = max(X)

minimum = min(X)
```

**Output**

```
maximum =

   30

minimum =

   12
```

**Explanation**

The functions max() returns the largest value, while min() returns the smallest value. These measures help identify the limits of the dataset.

### Range of a dataset

The range measures the spread of the data and is computed as the difference between the maximum and minimum values.

**Example:**

```
X = [12 15 18 21 24 18 15 30];

R = max(X) – min(X)
```

**Output**

```
R =

    18
```

**Explanation**

The range provides a simple measure of variability and indicates how widely the observations are distributed. A summary of descriptive statistics functions is provided in Table 9.3.

TABLE 9.3 Summary of descriptive statistics in MATLAB

| Functions | Purpose |
|---|---|
| **mean(X)** | Arithmetic average |
| **median(X)** | Middle value |
| **mode(X)** | Most frequent value |
| **max(X)** | Largest value |
| **min(X)** | Smallest value |
| **max(X)-min(X)** | Range of data |

These statistical measures provide an initial understanding of the dataset and are commonly used in engineering, scientific research, business analytics, and data-driven decision making.

## Measures of dispersion in MATLAB

Measures of central tendency include the mean, median, and mode. These measurements indicate the central tendency of a dataset but do not provide insights into the dispersion of the data values relative to the center. Dispersion measurements indicate the variability or distribution of the observations.

Low dispersion in a data collection indicates that data values are closely clustered, while high dispersion signifies that data values are more widely distributed. The variability of data is crucial in scientific research, engineering measurements, financial analysis, and statistical modeling.

### Variance of a dataset

Variance measures the average squared deviation of the observations from the mean. It provides an indication of how far the data values are spread from the average value.

MATLAB computes the sample variance using the var() function.

**Syntax**

```
var(X)
```

**Example:**

X = [12 15 18 21 24 18 15 30];

V = var(X)

**Output**

V =

34.1250

**Explanation**

A larger variance indicates greater variability in the dataset, whereas a smaller variance indicates that the observations are more closely clustered around the mean.

### Standard deviation

Standard deviation is one of the most commonly used measures of dispersion. It is defined as the square root of the variance and provides a measure of spread in the same units as the original data. MATLAB computes the standard deviation using the std() function.

**Syntax**

```
std(X)
```

**Example:**

```
X = [12 15 18 21 24 18 15 30];

S = std(X)
```

**Output**

```
S =

   5.8417
```

**Explanation**

The standard deviation estimates the average deviation of data from the mean:

– Minimal standard deviation → observations are closely grouped

– Elevated standard deviation (dispersed observations)

The standard deviation is typically more interpretable than variance, as it is conveyed in the same units as the data.

### Comparison of two datasets

Consider the datasets:

```
A = [18 19 20 21 22];
B = [5 10 20 30 35];
```

Both datasets have approximately the same mean, but their spreads are different.

**Example:**

```
std(A)

std(B)
```

**Output**

```
ans =

   1.5811

ans =

   12.7475
```

**Explanation**
The larger standard deviation of dataset B indicates greater variability compared to dataset A. This illustrates how standard deviation helps quantify the consistency of observations.

## Data sorting and organization in MATLAB

In computational and statistical analysis, data is often required to be arranged in a systematic order before further processing. Sorting and organizing data help identify trends, locate specific values, perform statistical analysis, and improve the interpretation of datasets.

MATLAB provides several built-in functions for sorting data and locating elements that satisfy specific conditions. These functions are particularly useful when working with large datasets in scientific computing, engineering analysis, and data-driven applications.

### Sorting data in MATLAB

The sort() function is used to arrange the elements of a vector or matrix in ascending order.

**Syntax**

```
sort(X)
```

**Example: Sorting in ascending order**
X = [25 12 30 18 10 22];

Y = sort(X)

**Output**
Y =

```
10   12   18   22   25   30
```

**Explanation**
The sort() function rearranges the elements of the vector in ascending order, placing the smallest value first and the largest value last.

### Sorting data in descending order

MATLAB also allows sorting data in descending order.

**Example:**

```
X = [25 12 30 18 10 22];

Y = sort(X,'descend')
```

**Output**

```
Y =

   30   25   22   18   12   10
```

**Explanation**

The option 'descend' instructs MATLAB to arrange the elements from largest to smallest.

### Sorting along matrix dimensions

When applied to a matrix, MATLAB sorts each column independently by default.

**Example:**

```
A = [9 4 7;
   3 8 2;
   6 1 5];

B = sort(A)
```

**Output**

```
B =

   3   1   2
   6   4   5
   9   8   7
```

**Explanation**

Each column is sorted separately while maintaining the matrix structure. This feature is useful for organizing multidimensional datasets.

### Finding the position of elements

MATLAB provides the find() function to determine the positions of elements that satisfy a specified condition.

**Syntax**

```
find(condition)
```

**Example:**
X = [12 18 25 30 15 22];

P = find(X > 20)

**Output**
P =

3 4 6

**Explanation**
The command find(X > 20) returns the positions of all elements greater than 20.

### Finding maximum and minimum positions

MATLAB can return both the value and its position using the functions max() and min().

**Example:**
X = [12 18 25 30 15 22];

[maxValue,indexMax] = max(X)

[minValue,indexMin] = min(X)

**Output**
maxValue =
30

indexMax =

4

minValue =

12

indexMin =

1

**Explanation**
The functions return the maximum (or minimum) value, and its corresponding location within the dataset.

### Data organization using logical conditions

MATLAB supports logical operations for selecting data that satisfy specific criteria.

**Example:**
X = [12 18 25 30 15 22];

Y = X(X > 20)

**Output**
Y =

25 30 22

**Explanation**
The expression X(X > 20) extracts only those values that satisfy the specified condition. This technique is known as logical indexing and is widely used for filtering datasets during data analysis.

A list of common data organization functions in MATLAB is provided in Table 9.4.

TABLE 9.4 Common data organization functions in MATLAB

| Function | Purpose |
|---|---|
| **sort(X)** | Sort data in ascending order |
| **sort(X,'descend')** | Sort data in descending order |
| **find(condition)** | Find positions satisfying a condition |

TABLE 9.4 Common data organization functions in MATLAB (*cont.*)

| Function | Purpose |
|---|---|
| **max(X)** | Largest value |
| **min(X)** | Smallest value |
| **X(condition)** | Extract values satisfying a condition |

## Working with Python

Python provides a versatile programming environment for numerical computation, scientific analysis, and data evaluation. Python possesses an extensive array of libraries that enable users to do mathematical computations, handle datasets, execute statistical analyses, and resolve computational issues efficiently. Robust tools like Math, NumPy, and Statistics facilitate the readability and adaptability of numerical and analytical activities.

One of the notable advantages of Python is its integration of general-purpose programming with advanced computational capabilities. Python is appropriate for several applications, including scientific research, engineering simulations, machine learning, and business analytics.

Python possesses comparable computational capabilities to MATLAB and may be broadly categorized into two primary domains:

**Numerical computation:** The examination of the efficacy of mathematical calculations and the resolution of computational issues

**Data analysis:** The utilization of statistical and analytical methodologies to extract valuable insights from numerical data collections

This section will examine several computational and data analysis capabilities available in Python, accompanied by practical applications.

## Numerical computation in Python

Numerical computation is an important part of scientific programming. This includes evaluating mathematical expressions, performing numerical calculations, solving equations and manipulating numerical datasets. Python has many built-in functions and special libraries that make these tasks easier and also allow for efficient computation.

The most popular libraries for numerical computation are:

| Library | Purpose |
| --- | --- |
| **math** | Mathematical functions |
| **numpy** | Numerical and scientific computing |
| **statistics** | Statistical computations |

Before using the functions available in these libraries, they must be imported into the program.

### Importing mathematical libraries

The math module provides functions for common mathematical operations, whereas NumPy offers advanced numerical capabilities for scientific computing.

**Example:**
import math
import numpy as np

**Explanation**
The statement:

**import math**
imports the mathematical functions available in Python's standard library.
Similarly,

**import numpy as np**
imports the NumPy library using the alias np, which is widely adopted in scientific computing.
These libraries provide efficient tools for performing numerical calculations and mathematical analysis.

### Mathematical computations using Python

The math module in Python offers an array of intrinsic mathematical functions. It allows users to perform complex numerical computations without the necessity of manually executing mathematical methods.

Table 9.5 lists many mathematical functions that are commonly employed in practical applications.

TABLE 9.5 List of mathematical computation commands in Python.

| Functions | Description |
|---|---|
| **math.sqrt(x)** | Square root |
| **math.fabs(x)** | Absolute value |
| **math.exp(x)** | Exponential function |
| **math.log(x)** | Natural logarithm |
| **math.log10(x)** | Base-10 logarithm |
| **math.sin(x)** | Sine function |
| **math.cos(x)** | Cosine function |
| **math.tan(x)** | Tangent function |
| **round(x)** | Round to nearest integer |
| **math.floor(x)** | Round downward |
| **math.ceil(x)** | Round upward |

**Example: Evaluating mathematical functions**

```
import math
x = 16
a = math.sqrt(x)
b = math.fabs(-25)
c = math.exp(2)
d = math.log(5)
e = round(4.67)
print(a,b,c,d,e)
```

**Output**

```
4.0 25.0 7.38905609893065 1.6094379124341003 5
```

**Explanation**

The math module provides optimized implementations of commonly used mathematical functions. These functions are extensively used in scientific computing, engineering calculations, and numerical modeling.

### Trigonometric and exponential computations in Python

Trigonometric and exponential functions are frequently employed in scientific computing, engineering analysis, signal processing, machine learning, and mathematical modeling. The Python math module includes functions for precise and efficient numerical computations.

The trigonometric functions in Python operate using angles measured in radians. Consequently, the angles specified in degrees must be translated to radians before to calculations. The commonly used trigonometric functions of Python are listed in Table 9.6.

TABLE 9.6 Common trigonometric functions of Python.

| Function | Description |
|---|---|
| **math.sin(x)** | Sine function |
| **math.cos(x)** | Cosine function |
| **math.tan(x)** | Tangent function |
| **math.asin(x)** | Inverse sine |
| **math.acos(x)** | Inverse cosine |
| **math.atan(x)** | Inverse tangent |

**Example: Trigonometric computations**

```
import math
x = math.pi/4
s = math.sin(x)
c = math.cos(x)
t = math.tan(x)
print(s)
print(c)
print(t)
```

**Output**

```
0.7071
0.7071
1.0000
```

**Explanation**

The value:

x = math.pi/4

represents an angle of $45^\circ$ in radians.

Python computes the sine of $45^\circ$, the cosine of $45^\circ$, and the tangent of $45^\circ$. These functions are commonly used in wave analysis, electrical engineering, computer graphics, and scientific simulations.

### Exponential and logarithmic functions

Exponential and logarithmic computations are important in growth models, population studies, finance, machine learning, and engineering applications.

Python provides built-in functions for these computations through the math module.

**Example:**

```
import math
x = 5
a = math.exp(x)
b = math.log(x)
c = math.log10(x)
print(a)
print(b)
print(c)
```

**Output**

```
148.4132
1.6094
0.6990
```

**Explanation**

In this example, math.exp(x) computes $e^x$, math.log(x) computes the natural logarithm, and math.log10(x) computes the logarithm to base 10.

These functions are frequently used in scientific computations and mathematical modeling.

### Complex number computations in Python

Complex numbers are frequently employed in electrical engineering, signal processing, control systems, classical quantum mechanics, and scientific computations. Python has built-in assistance for complex arithmetic, no more specialized libraries needed.

A complex number is represented in the form:

$$z = a + bj$$

where *a* is the real part, *b* is the imaginary part, and *j* represents the imaginary unit.

Unlike MATLAB, Python uses *j* instead of *i* for representing imaginary numbers.

### Defining complex numbers

Complex numbers can be defined directly using the imaginary unit j.

**Example:**

```
z1 = 3 + 4j
z2 = 2 – 5j
print(z1)
print(z2)
```

**Output**

```
(3+4j)
(2–5j)
```

### Basic operations on complex numbers

Python supports arithmetic operations on complex numbers in the same way as real numbers.

**Example:**

```
z1 = 3 + 4j
z2 = 2 – 5j
sum_z = z1 + z2
prod_z = z1 * z2
print(sum_z)
print(prod_z)
```

**Output**

```
(5–1j)
(26–7j)
```

### Extracting components of a complex number

Python provides built-in attributes and functions for accessing different properties of complex numbers.

**Example:**

```
z = 3 + 4j
print(z.real)
print(z.imag)
print(abs(z))
print(z.conjugate())
```

**Output**

```
3.0
4.0
5.0
(3-4j)
```

**Explanation**
- z.real returns the real part,
- z.imag returns the imaginary part,
- abs(z) computes the magnitude, and
- z.conjugate() computes the complex conjugate.

Complex arithmetic forms the foundation of many advanced engineering and scientific applications.

### Polynomial computations in Python

Polynomials are extremely prevalent in engineering, mathematics, physics, control systems and numerical analysis. NumPy has a few very strong polynomial manipulation capabilities in Python. These functions provide users with the ability to define, evaluate, and analyze polynomial expressions efficiently.

A polynomial of degree $n$ can be expressed as:

$$P(x) = a_n x^n + a_{n-1} x^{n-1} + \cdots + a_1 x + a_0$$

Similar to MATLAB, NumPy represents a polynomial using a vector containing its coefficients arranged in descending powers of the variable.

#### Defining a polynomial

Consider the polynomial:

$$P(x) = x^3 - 6x^2 + 11x - 6$$

The coefficient vector is defined as:

```
import numpy as np
P = [1, -6, 11, -6]
```

The coefficients correspond to:

$$1x^3 - 6x^2 + 11x - 6$$

#### Evaluating a polynomial

Python provides the numpy.polyval() function to evaluate a polynomial at a specified value.

**Syntax**

```
np.polyval(P,x)
```

**Example:**

```
import numpy as np
P = [1,-6,11,-6]
value = np.polyval(P,2)
print(value)
```

**Output**

```
0
```

**Explanation**

The command evaluates:

$$P(2) = 2^3 - 6(2)^2 + 11(2) - 6$$

which results in zero.

### Finding polynomial roots

The roots of a polynomial are the values of $x$ for which:

$$P(x) = 0$$

Python provides the numpy.roots() function for determining polynomial roots.

**Example:**

```
import numpy as np
P = [1,−6,11,−6]
r = np.roots(P)
print(r)
```

**Output**

```
[3. 2. 1.]
```

**Explanation**

The polynomial has roots at:

$$x = 1, x = 2, x = 3$$

The roots() function is widely used in engineering computations, numerical methods, and control system analysis.

### Constructing a polynomial from roots

If the roots of a polynomial are known, Python can generate the corresponding polynomial coefficients.

**Example:**

```
import numpy as np
r = [1,2,3]
P = np.poly(r)
print(P)
```

**Output**

```
[ 1. -6. 11. -6.]
```

**Explanation**

The function np.poly() generates the coefficient vector of the polynomial having the specified roots.

### Numerical integration and differentiation in Python

Numerical differentiation and integration are important computational techniques used in engineering, physics, scientific computing, and applied mathematics. Python provides several methods for approximating derivatives and integrals when analytical solutions are difficult to obtain.

These techniques are particularly useful when working with discrete datasets obtained from experiments or simulations.

### Numerical differentiation

Python provides the diff() function in NumPy for computing differences between consecutive elements.

**Example:**

```
import numpy as np
x = np.array([1,2,4,7,11])
d = np.diff(x)
print(d)
```

**Output**

[1 2 3 4]

**Explanation**

The command computes:

$$[2-1, |4-2, |7-4, |11-7]$$

which represents the differences between consecutive elements.
The diff() function is often used to approximate derivatives and rates of change.

### Numerical integration using the trapezoidal rule

NumPy provides the trapz() function for numerical integration using the trapezoidal rule.

**Example:**

```
import numpy as np
x = np.arange(0,2.5,0.5)
y = x**2
I = np.trapz(y,x)
print(I)
```

**Output**

2.75

**Explanation**

The command np.trapz(y,x) approximates the area under the curve $y = x^2$ over the interval $[0, 2]$.

## Data analysis in Python

Data analysis is the procedure of analyzing, structuring, summarizing, and interpreting data to find meaningful information. In mathematical engineering and computation applications, data analysis can be used to detect trends, patterns, and variations, as well as connections within datasets. Python has a number of built-in libraries for statistical analysis and numerical data processing.

The most common libraries used for basic data analysis are:

- statistics and
- numpy

They provide efficient functions to calculate measures of central tendency and dispersion.

### Descriptive statistical measures

Descriptive statistics summarize the important characteristics of a dataset using numerical measures. These measures help understand the central tendency and distribution of the data.

Consider the following dataset:

X = [12,15,18,21,24,18,15,30]

#### Mean of a dataset

The mean represents the arithmetic average of all observations in a dataset.

**Syntax**

```
statistics.mean(X)
```

**Example:**

```
import statistics
X = [12,15,18,21,24,18,15,30]
M = statistics.mean(X)
print(M)
```

**Output**

19.125

**Explanation**

The mean is calculated by dividing the sum of all observations by the total number of observations. The mean provides a measure of the central location of the dataset.

#### Median of a dataset

The median represents the middle value when the observations are arranged in ascending order.

**Syntax**

```
statistics.median(X)
```

**Example:**

```
import statistics
X = [12,15,18,21,24,18,15,30]
Md = statistics.median(X)
print(Md)
```

**Output**

18.0

**Explanation**

The median is less affected by extreme observations and therefore provides a robust measure of central tendency.

### Mode of a dataset

The mode represents the value that occurs most frequently in a dataset.

**Syntax**

```
statistics.mode(X)
```

**Example:**

```
import statistics
X = [12,15,18,21,24,18,15,30]
Mo = statistics.mode(X)
print(Mo)
```

**Output**

15

**Explanation**

The mode identifies the most frequently occurring observation in the dataset.

### Maximum and minimum values

The largest and smallest observations provide useful information regarding the limits of a dataset.

**Example:**

```
X = [12,15,18,21,24,18,15,30]
maximum = max(X)
minimum = min(X)
print(maximum)
print(minimum)
```

**Output**

```
30
12
```

**Explanation**

The functions max() return the largest value, and min() return the smallest value. These measures help identify the range of the dataset.

### Range of a dataset

The range is the difference between the maximum and minimum values.

**Example:**

```
X = [12,15,18,21,24,18,15,30]
R = max(X) – min(X)
print(R)
```

**Output**

```
18
```

**Explanation**

The range provides a simple measure of variability and indicates how widely the observations are distributed.

A summary of functions for the descriptive statistics is listed in Table 9.7 mentioning the function along with the details.

TABLE 9.7 Summary of descriptive statistics in Python

| Functions | Purpose |
|---|---|
| **statistics.mean(X)** | Arithmetic average |
| **statistics.median(X)** | Middle value |
| **statistics.mode(X)** | Most frequent value |
| **max(X)** | Largest value |
| **min(X)** | Smallest value |
| **max(X)-min(X)** | Range of data |

These statistical measures provide a preliminary understanding of the dataset and are widely used in scientific research, engineering analysis, business analytics, and data science applications.

### Measures of dispersion in Python

While the mean, median, and mode describe the central tendency of a dataset, measures of dispersion describe how widely the observations are distributed around the center.

Python provides built-in statistical functions for computing:

- variance,
- standard deviation, and
- other measures of variability.

These measures help assess the consistency and spread of data.

#### Variance of a dataset

Variance measures the average squared deviation of the observations from the mean.

**Syntax**

```
statistics.variance(X)
```

**Example:**

```
import statistics
X = [12,15,18,21,24,18,15,30]
V = statistics.variance(X)
print(V)
```

**Output**

34.125

**Explanation**

A larger variance indicates greater variability in the dataset, whereas a smaller variance indicates that the observations are more closely clustered around the mean.

### Standard deviation

The standard deviation is the square root of the variance and is one of the most commonly used measures of dispersion.

**Syntax**

```
statistics.stdev(X)
```

**Example:**

```
import statistics
X = [12,15,18,21,24,18,15,30]
S = statistics.stdev(X)
print(S)
```

**Output**

5.8417

**Explanation**

The standard deviation measures the average distance of observations from the mean:

- Small standard deviation → observations are closely grouped.
- Large standard deviation → observations are widely spread.

Because it is expressed in the same units as the data, standard deviation is often easier to interpret than variance.

## Data sorting and organization in Python

In computational and statistical analysis, data are often required to be sorted in a systematic order prior to further processing. Data sorting and organization helps in trend recognition, finding specific values, statistical analysis and better interpretation of datasets. Python has several built-in functions and NumPy utilities to sort data and find elements that satisfy certain criteria. These tools are especially important for working with large datasets in scientific computing, engineering analysis, machine learning and data science applications.

### Sorting data in Python

Python provides the sorted() function for arranging data in ascending order.

**Syntax**

```
sorted(X)
```

**Example: Sorting in ascending order**

```
X = [25,12,30,18,10,22]
Y = sorted(X)
print(Y)
```

**Output**

```
[10, 12, 18, 22, 25, 30]
```

**Explanation**
The sorted() function returns a new list containing the elements arranged in ascending order. The original list remains unchanged.

#### Sorting data in descending order

Python allows sorting data in descending order using the reverse=True option.

**Example:**

```
X = [25,12,30,18,10,22]
Y = sorted(X, reverse=True)
print(Y)
```

**Output**
[30, 25, 22, 18, 12, 10]

**Explanation**
The argument:
reverse=True
instructs Python to arrange the elements from largest to smallest.

### Sorting NumPy arrays

When working with numerical data, NumPy provides the sort() function.

**Example:**
```
import numpy as np
X = np.array([25,12,30,18,10,22])
Y = np.sort(X)
print(Y)
```

**Output**
[10 12 18 22 25 30]

**Explanation**
The np.sort() function sorts the elements of a NumPy array and returns the sorted array.

### Finding the position of elements

NumPy provides the where() function to determine the positions of elements satisfying a specified condition.

**Syntax**

```
np.where(condition)
```

**Example:**
```
import numpy as np
X = np.array([12,18,25,30,15,22])
P = np.where(X > 20)
print(P)
```

**Output**
(array([2, 3, 5]),)

**Explanation**
The command:
np.where(X > 20)
returns the indices of all elements greater than 20.
Since Python uses zero-based indexing:
- 25 occurs at index 2,
- 30 occurs at index 3, and
- 22 occurs at index 5.

### Finding maximum and minimum positions

NumPy provides the functions argmax() and argmin() to determine the positions of the largest and smallest elements.

**Example:**
```
import numpy as np
X = np.array([12,18,25,30,15,22])
maximum = np.max(X)
minimum = np.min(X)
indexMax = np.argmax(X)
indexMin = np.argmin(X)
print(maximum)
print(indexMax)
print(minimum)
print(indexMin)
```

**Output**
30
3
12
0

**Explanation**
The functions:
- np.max() return the maximum value,
- np.min() return the minimum value,
- np.argmax() return the index of the maximum value, and
- np.argmin() return the index of the minimum value.

These functions are useful in optimization problems and data analysis applications.

### Data organization using logical conditions

Python supports logical indexing through NumPy arrays, allowing data to be filtered based on specified conditions.

**Example:**

```
import numpy as np
X = np.array([12,18,25,30,15,22])
Y = X[X > 20]
print(Y)
```

**Output**

```
[25 30 22]
```

**Explanation**

The expression:

X[X > 20]

extracts only those values that satisfy the specified condition.

This technique is known as **boolean indexing** and is widely used for filtering datasets during data analysis and machine learning applications.

The common data organization functions in Python are listed in Table 9.8 detailing the commonly used functions with details in Python.

TABLE 9.8 Common data organization functions in Python

| Functions | Purpose |
|---|---|
| **sorted(X)** | Sort data in ascending order |
| **sorted(X, reverse=True)** | Sort data in descending order |
| **np.sort(X)** | Sort NumPy arrays |
| **np.where(condition)** | Find indices satisfying a condition |
| **np.max(X)** | Largest value |
| **np.min(X)** | Smallest value |
| **np.argmax(X)** | Position of largest value |
| **np.argmin(X)** | Position of smallest value |
| **X[condition]** | Extract values satisfying a condition |

### Summary of Python data analysis

Python provides a rich collection of built-in functions and scientific libraries for performing data analysis efficiently. Functions available through the statistics and NumPy libraries allow users to compute descriptive statistics, analyze variability, organize datasets, and extract meaningful information from numerical data.

These tools form an important foundation for scientific computing, machine learning, artificial intelligence, and modern data-driven applications.

## MATLAB and Python: a comparative perspective

MATLAB and Python are among the most prevalent programming environments utilized in scientific computing, numerical analysis, and data processing. Both platforms provide comprehensive computing capabilities and are highly suitable for many mathematical, statistical, and engineering applications. MATLAB is a specific environment designed for numerical computing and matrix operations, whereas Python is a general-purpose programming language that offers an extensive array of scientific libraries.

The selection between Python and MATLAB is contingent upon the application's characteristics, available resources, and user requirements. MATLAB provides a cohesive environment and numerous built-in functions for calculation. Python use specific libraries such as NumPy, SciPy, Matplotlib, and Statistics to achieve similar functionality.

Table 9.9 compares frequently utilized computational operations in MATLAB and Python

TABLE 9.9 Common computational operations in MATLAB and Python.

| Operations | MATLAB | Python |
|---|---|---|
| **Square root** | sqrt(x) | math.sqrt(x) |
| **Absolute value** | abs(x) | abs(x) |
| **Exponential function** | exp(x) | math.exp(x) |
| **Natural logarithm** | log(x) | math.log(x) |
| **Sine function** | sin(x) | math.sin(x) |
| **Mean** | mean(X) | statistics.mean(X) |
| **Median** | median(X) | statistics.median(X) |
| **Mode** | mode(X) | statistics.mode(X) |
| **Variance** | var(X) | statistics.variance(X) |
| **Standard deviation** | std(X) | statistics.stdev(X) |

TABLE 9.9 Common computational operations in MATLAB and Python (*cont.*)

| Operations | MATLAB | Python |
|---|---|---|
| **Sort data** | sort(X) | sorted(X) |
| **Polynomial roots** | roots(P) | np.roots(P) |
| **Numerical integration** | trapz(x,y) | np.trapz(y,x) |

## Chapter summary

The basic concepts of computational mathematics and data analysis with MATLAB and Python have been presented in this chapter. We explored a variety of techniques for numerical calculations, evaluating mathematical expressions, and solving computational problems using the tools and libraries inherent to both programming environments.

In the chapter, we have seen how to do mathematical functions, trigonometric and exponential calculations, arithmetic operations with complex numbers, polynomial calculations, and numerical integration and differentiation. The presentation also included measures of central tendency, measures of dispersion and data organization techniques for statistical analysis. These tools allow programmers to efficiently handle numerical data and extract useful insights from datasets.

This comparison discussed the similarities and differences between MATLAB and Python and helped the reader see how to perform the same computational tasks in each environment.

The ideas presented in this chapter provide an essential background for scientific computing, engineering analysis, machine learning, data science, and many other computational applications.

## Exercise questions

### Section A: MATLAB

1. Write a MATLAB program to evaluate:

$$y = \sqrt{x} + x^2$$

for $x = 10$.

2. Write a MATLAB program to compute the sine, cosine, and tangent of $45^\circ$.
3. Create a MATLAB program to perform arithmetic operations on two complex numbers.
4. Evaluate the polynomial:

$$P(x) = x^3 - 4x^2 + 5x - 2$$

at $x = 2$.
5. Determine the roots of the polynomial:

$$x^3 - 6x^2 + 11x - 6 = 0$$

6. Use diff() to determine the differences between consecutive elements of a vector.
7 Use trapz() to estimate the area under the curve:

$$y = x^2$$

8. Compute the mean, median, and mode of a given dataset.
9. Determine the variance and standard deviation of a dataset.
10. Sort a dataset in ascending and descending order.

### Section B: Python

1. Write a Python program to evaluate:

$$y = \sqrt{x} + x^2$$

for $x = 10$.
2. Write a Python program to compute the sine, cosine, and tangent of $45^\circ$.
3. Create a Python program to perform arithmetic operations on two complex numbers.
4. Evaluate the polynomial:

$$P(x) = x^3 - 4x^2 + 5x - 2$$

at $x = 2$.
5. Determine the roots of the polynomial:

$$x^3 - 6x^2 + 11x - 6 = 0$$

6. Use np.diff() to determine differences between consecutive elements.
7. Use np.trapz() to estimate the area under a curve.
8. Compute the mean, median, and mode of a dataset.
9. Determine the variance and standard deviation of a dataset.
10. Filter all values greater than 20 from a NumPy array using boolean indexing.

## Computational practice questions with short answers

1. **Data cleaning and missing values**
   **Question:** How can missing values in a dataset be identified?
   **Short answer:** Use functions to locate NaN or missing entries.

**MATLAB**

```
A = [12 NaN 18 25];
isnan(A)
```

**Python**

```
import pandas as pd
df = pd.DataFrame({'A':[12,None,18,25]})
df.isnull()
```

2. **Descriptive statistics**
   **Question:** Compute the mean, median, and standard deviation of a dataset.
   **Short answer:** Mean measures central tendency, median gives middle value, and standard deviation measures dispersion.

**MATLAB**

```
x = [10 15 20 25 30];
mean(x), median(x), std(x)
```

**Python**

```
import numpy as np
x = [10,15,20,25,30]
np.mean(x), np.median(x), np.std(x,ddof=1)
```

3. **Frequency distribution**
   **Question:** Determine the frequency of each unique value in a dataset.
   **Short answer:** Frequency table summarizes data occurrences.

**MATLAB**

```
x = [1 2 2 3 3 3];
tabulate(x)
```

**Python**

```
import pandas as pd
pd.Series([1,2,2,3,3,3]).value_counts()
```

4. **Data normalization**
   **Question:** Normalize data between 0 and 1.
   **Short answer:** Scale values using Min-Max normalization.

**MATLAB**

```
x = [10 20 30 40];
(x-min(x))/(max(x)-min(x))
```

**Python**

```
import numpy as np
x=np.array([10,20,30,40])
(x-x.min())/(x.max()-x.min())
```

5. **Correlation analysis**
   **Question:** Find the correlation between study hours and exam scores.
   **Short answer:** Correlation coefficient indicates relationship strength.

**MATLAB**

```
x=[2 4 6 8];
y=[40 55 70 85];
corrcoef(x,y)
```

**Python**

```
import numpy as np
np.corrcoef([2,4,6,8],[40,55,70,85])
```

6. **Linear regression**
   **Question:** Fit a linear regression model to a dataset.
   **Short answer:** Regression predicts dependent variables from independent variables.

**MATLAB**

```
x=[1 2 3 4];
y=[2 4 6 8];
polyfit(x,y,1)
```

**Python**

```
import numpy as np
np.polyfit([1,2,3,4],[2,4,6,8],1)
```

7. **Data visualization**
   **Question:** Plot monthly sales data.
   **Short answer:** Line plots reveal trends over time.

**MATLAB**

```
sales=[100 120 150 170];
plot(sales)
```

**Python**

```
import matplotlib.pyplot as plt
plt.plot([100,120,150,170])
plt.show()
```

### Mini project: student performance analytics using MATLAB and Python

Project title: computational analysis of student academic performance using MATLAB and Python

**Dataset:** The following data were collected from five students.

| Student | Study hours (per week) | Attendance (%) | Assignment marks | Final marks |
|---|---|---|---|---|
| 1 | 5 | 70 | 60 | 58 |
| 2 | 8 | 78 | 70 | 71 |
| 3 | 11 | 84 | 78 | 80 |
| 4 | 14 | 88 | 84 | 87 |
| 5 | 17 | 92 | 88 | 92 |

- **Task 1:** Import the dataset into MATLAB and Python.
- **Task 2:** Calculate:
  - Mean
  - Median
  - Standard deviation
  - Variance

  for all variables.
- **Task 3:** Create the following visualizations:
  - Histogram of final marks
  - Box plot of study hours
  - Scatter plot of study hours versus final marks
- **Task 4:** Compute the correlation matrix.
- **Task 5:** Determine which variable has the strongest relationship with final marks.

CHAPTER 10

# Symbolic computation

Symbolic computation (also called computer algebra) is the manipulation and evaluation of mathematical expressions in their exact analytical form rather than numerical approximations. Numerical computation yields decimal values. Symbolic computation retains variables, constants, fractions, equations, functions, etc. in their mathematical form. This allows mathematical expressions to be simplified, differentiated, integrated, factored and solved with exact results.

Symbolic mathematics is important in scientific computing, engineering analysis, optimization, control systems, differential equations, and mathematical modeling. Symbolic computation retains analytical expressions rather than numerical approximations, providing a deeper understanding of mathematical relationships and resulting in more precise solutions.

For example:

| Numerical computation | Symbolic computation |
|---|---|
| $\sqrt{2} \approx 1.4142$ | $\sqrt{2}$ |
| $1/3 \approx 0.3333$ | $1/3$ |

Current computing systems have special software for symbolic maths. MATLAB has a symbolic computation toolbox called Symbolic Math Toolbox. You have a powerful symbolic computation library called SymPy in Python. These tools allow users to carry out algebraic simplification, symbolic differentiation, integration, equation solving, series expansions, matrix operations, and many other mathematical tasks efficiently.

This chapter introduces the basic concepts of symbolic computation with the practical examples and the comparative discussion of different symbolic mathematics techniques such as MATLAB and Python.

## Working with MATLAB

MATLAB can perform symbolic computations with the Symbolic Math Toolbox. This toolbox enables you to symbolically represent and manipulate mathematical expressions, rather than numerically. The symbolic variables enable the users to

 | HTTPS://DOI.ORG/10.1515/9783112207758-010

perform exact calculations, to simplify algebraic expressions, to solve the equations analytically, to calculate the derivatives and integrals, and to get the closed-form solutions of the mathematical problems.

Another major advantage of symbolic computation is that it is mathematically exact and thus free from rounding and approximation errors, which are often present in numerical computations. Thus, symbolic methods are extensively used in engineering analysis, scientific research, optimization problems, control systems and mathematical modeling.

The Symbolic Math Toolbox offers a vast array of functions for symbolic manipulation, allowing users to execute complicated math operations using simple commands. In the following sections, we shall take a look at a few of the symbolic computation techniques available in MATLAB with practical examples.

The first step in symbolic computation in MATLAB is to define symbolic variables. Symbolic variables are different from ordinary numeric variables in that they represent mathematical quantities that can be manipulated algebraically. These variables are used as a basis for symbolic calculations such as simplification, solving equations, differentiation, integration, and series expansion.

In MATLAB, you define symbolic variables using the syms command. Symbolic variables can be defined, and then used to construct mathematical expressions and to perform exact calculations.

## Creating symbolic variables

Before performing symbolic computations, symbolic variables must be declared. MATLAB uses the syms command to create symbolic variables that can be manipulated algebraically.

**Example: Creating symbolic variables**

```
syms x y
```

**Output:**

```
x
y
```

**Explanation:**

The command syms x y, creates two symbolic variables, x and y. Unlike ordinary variables that store numerical values, symbolic variables retain their algebraic identity and can be used directly in mathematical expressions.

For example:

```
syms x
f = x^2 + 3*x + 2
```

produces:

```
f =
x^2 + 3*x + 2
```

The expression is left in symbolic form and then can be simplified, differentiated, integrated, or solved analytically.

All symbolic calculations in MATLAB are based on symbolic variables and they are used extensively in algebraic manipulation, calculus, and equation solving.

## Expression manipulation

MATLAB provides various symbolic functions for manipulating algebraic expressions. The simplify() function reduces an expression to its simplest form, expand() converts factored expressions into polynomial form, and factor() decomposes polynomials into simpler factors.

**Example:**

```
syms x
simplify((x+1)^2-(x^2+2*x+1))
expand((x+2)^3)
factor(x^2-9)
```

**Output:**

```
0
x^3 + 6*x^2 + 12*x + 8
(x-3)*(x+3)
```

These operations are useful in algebraic simplification, polynomial analysis, and symbolic equation solving.

## Solving symbolic equations

One of the most important applications of symbolic computation is obtaining exact solutions of algebraic equations. MATLAB provides the solve() function for determining symbolic solutions of equations involving one or more variables.

### Solving a single equation

Consider:

$$x^2 - 5x + 6 = 0$$

```
syms x
solve(x^2-5*x+6==0,x)
```

**Output:**

```
2
3
```

The solutions indicate that the equation has roots at $x = 2$ and $x = 3$.

### Solving a system of equations

Consider:

$$x + y = 5$$

$$x - y = 1$$

```
syms x y
solve([x+y==5, x-y==1],[x,y])
```

**Output:**

```
x = 3
y = 2
```

Symbolic equation solving is widely used in engineering analysis, optimization, and mathematical modeling.

## Symbolic calculus

Calculus operations such as differentiation and integration can be performed symbolically using MATLAB. Symbolic calculus produces exact analytical results and eliminates approximation errors.

### Differentiation

MATLAB can also compute higher-order derivatives using command diff().

Consider:

$$f(x) = x^3 + 2x^2 + 5x$$

```
syms x
f = x^3 + 2*x^2 + 5*x;
diff(f,x)
```

**Output:**

```
3*x^2 + 4*x + 5
```

Example:

```
diff(f,x,2)
```

**Output:**

```
6*x + 4
```

### Integration

Consider:

$$\int x^2 \, dx$$

```
syms x
int(x^2,x)
```

**Output**

```
x^3/3
```

Definite integrals can also be evaluated directly:

```
int(x^2,x,0,1)
```

**Output**

```
1/3
```

Symbolic differentiation and integration are extensively used in engineering analysis, optimization, and scientific computing.

### Series expansion and limits

MATLAB provides functions for obtaining Taylor series approximations and evaluating mathematical limits symbolically.

#### Taylor series expansion

Consider the function $e^x$

```
syms x
taylor(exp(x))
```

**Output:**

1 + x + x^2/2 + x^3/6 + ...

The Taylor series approximates a function using polynomial terms and is widely used in numerical analysis and engineering applications.

#### Evaluating limits

Consider:

$$\lim_{x \to 0} \frac{\sin x}{x}$$

```
syms x
limit(sin(x)/x,x,0)
```

**Output:**

1

Symbolic limits are useful for analyzing the behavior of functions near singularities and critical points.

## Working with Python

Python provides symbolic computation capabilities through the **SymPy** library, an open-source package designed for symbolic mathematics. SymPy enables users to manipulate algebraic expressions, solve equations, perform differentiation and integration, evaluate limits, and generate series expansions while maintaining exact mathematical representations.

Unlike numerical libraries that produce approximate decimal values, SymPy preserves mathematical expressions in symbolic form, making it suitable for analytical problem-solving, engineering computations, and mathematical modeling.

### Creating symbolic variables

Symbolic computations in Python begin by creating symbolic variables using the symbols() function available in the SymPy library.

**Example:**

```
from sympy import *
x, y = symbols('x y')
f = x**2 + 3*x + 2
print(f)
```

**Output:**

```
x**2 + 3*x + 2
```

The variables remain symbolic and can be used for algebraic manipulations, differentiation, integration, and equation solving.

### Expression manipulation

SymPy provides several functions for manipulating algebraic expressions. The simplify() function reduces an expression to its simplest form, expand() converts factored expressions into polynomial form, and factor() decomposes polynomials into simpler factors.

**Example:**

```
from sympy import *
x = symbols('x')
simplify((x+1)**2-(x**2+2*x+1))
expand((x+2)**3)
factor(x**2-9)
```

**Output:**

```
0
x**3 + 6*x**2 + 12*x + 8
(x – 3)*(x + 3)
```

These operations are useful in algebraic simplification, polynomial analysis, and symbolic equation solving.

## Solving symbolic equations

SymPy provides the solve() function for obtaining exact solutions of algebraic equations.

### Solving a single equation

Consider:

$$x^2 - 5x + 6 = 0$$

```
from sympy import *
x = symbols('x')
solve(x**2-5*x+6,x)
```

**Output:**

[2, 3]

The solutions indicate that the equation has roots at $x = 2$ and $x = 3$.

### Solving a system of equations

Consider:

$$x + y = 5$$

$$x - y = 1$$

```
from sympy import *
x, y = symbols('x y')
solve((x+y-5, x-y-1),(x,y))
```

**Output:**

{x: 3, y: 2}

SymPy can efficiently solve systems of equations involving multiple variables.

## Symbolic calculus

SymPy provides powerful tools for symbolic differentiation and integration.

### Differentiation

Consider:

$$f(x) = x^3 + 2x^2 + 5x$$

```
from sympy import *
x = symbols('x')
diff(x**3 + 2*x**2 + 5*x, x)
```

**Output:**
3*x**2 + 4*x + 5
Higher-order derivatives can also be computed:
diff(x**3 + 2*x**2 + 5*x, x, 2)

**Output:**
6*x + 4

### Integration

Consider:

$$\int x^2 \, dx$$

```
from sympy import *
x = symbols('x')
integrate(x**2,x)
```

**Output:**
x**3/3
Definite integrals can be evaluated directly:
integrate(x**2,(x,0,1))

**Output:**
1/3
Symbolic differentiation and integration enable exact analytical solutions without numerical approximations.

## Series expansion and limits

SymPy provides functions for generating Taylor series expansions and evaluating limits symbolically.

### Taylor series expansion

Consider:

$$e^x$$

```
from sympy import *
x = symbols('x')
series(exp(x),x,0,6)
```

**Output:**

```
1 + x + x**2/2 + x**3/6 + x**4/24 + x**5/120
```

The Taylor series approximates a function using polynomial terms and is widely used in numerical analysis and engineering applications.

### Evaluating limits

Consider:

$$\lim_{x \to 0} \frac{\sin(x)}{x}$$

```
from sympy import *
x = symbols('x')
limit(sin(x)/x,x,0)
```

**Output:**

```
1
```

Symbolic limits help analyze the behavior of functions near singularities and critical points.

## MATLAB and Python: a comparative perspective

MATLAB and Python are preferred platforms for symbolic computation and for mathematical analysis. Both systems allow users to manipulate algebraic expressions, solve equations, perform symbolic differentiation and integration, evaluate limits and generate series expansions. MATLAB has symbolic capabilities with the Symbolic Math Toolbox, and Python has them with the open-source SymPy library. The syntax is slightly different but the symbolic operations are quite similar.

Table 10.1 shows a comparison of commonly used symbolic computation commands in MATLAB and Python.

TABLE 10.1 Commonly used symbolic computation commands in MATLAB and Python.

| Operations | MATLAB | Python (SymPy) |
|---|---|---|
| **Create symbolic variable** | syms x | x = symbols('x') |
| **Simplify expression** | simplify(expr) | simplify(expr) |
| **Expand expression** | expand(expr) | expand(expr) |
| **Factorize expression** | factor(expr) | factor(expr) |
| **Solve equation** | solve(eqn,x) | solve(eqn,x) |
| **Solve system of equations** | solve([eqns],[vars]) | solve((eqns),(vars)) |
| **Differentiate** | diff(f,x) | diff(f,x) |
| **Integrate** | int(f,x) | integrate(f,x) |
| **Definite integral** | int(f,x,a,b) | integrate(f,(x,a,b)) |
| **Taylor series** | taylor(f) | series(f,x,o,n) |
| **Evaluate limit** | limit(f,x,a) | limit(f,x,a) |

MATLAB offers a highly integrated environment with built-in symbolic capabilities, making it very convenient for engineering and scientific applications. SymPy in Python provides similar symbolic capabilities in an open-source environment and can be easily combined with other scientific computing libraries. Both platforms are capable of solving a large variety of symbolic mathematics problems and are widely used in research, education, and industrial applications.

## Chapter summary

Symbolic computation is the manipulation of mathematical expressions in their analytical form instead of numerical approximations. Symbolic methods do not lose any information. In contrast to numerical computation, symbolic methods maintain variables, constants, fractions, and functions.

Mathematical operations can be done with exact precision. This chapter presented symbolic computation with MATLAB and Python. Practical examples were used to demonstrate the process of creating symbolic variables, simplifying expressions, expanding, factoring, solving equations, symbolic differentiation, integration, series expansion, and evaluating limits. Symbolic math operations: MATLAB has Symbolic Math Toolbox for symbolic operations. Python has the SymPy library that provides similar functionality. In the comparative discussion, the similarities of MATLAB and Python for symbolic mathematics were highlighted and it was shown how the equivalent operation can be performed in the two environments. The techniques discussed in this chapter form an important foundation for advanced mathematical modeling, optimization, engineering analysis, scientific computing, and computer algebra systems.

## Exercise questions

### Section A: MATLAB

1. Create symbolic variables $x$ and $y$ and define the expression:

$$f(x) = x^2 + 4x + 3$$

2. Simplify the expression:

$$(x+2)^2 - (x^2 + 4x + 4)$$

3. Expand:

$$(x+1)^4$$

4. Factorize:

$$x^2 - 16$$

5. Solve the equation:

$$x^2 - 7x + 12 = 0$$

6. Solve the system:

$$x + y = 10$$

$$x - y = 4$$

7. Determine:

$$\frac{d}{dx}(x^4 + 2x^3 + 5x)$$

8. Evaluate:

$$\int x^3 \, dx$$

9. Obtain the Taylor series expansion of:

$$\sin(x)$$

10. Evaluate:

$$\lim_{x \to 0} \frac{1 - \cos(x)}{x^2}$$

## Section B: Python

1. Create symbolic variables $x$ and $y$ using SymPy.
2. Simplify the expression:

$$\frac{x^2 - 1}{x - 1}$$

3. Expand:

$$(x + 3)^3$$

4. Factorize:

$$x^2 - 25$$

5 Solve:

$$x^2 - 9x + 20 = 0$$

6. Solve the system:

$$x + y = 6$$

$$x - y = 2$$

7. Determine:

$$\frac{d}{dx}(x^3 + 3x^2 + 1)$$

8. Evaluate:

$$\int x^2 \, dx$$

9. Generate the Taylor series expansion of:

$$e^x$$

10. Evaluate:

$$\lim_{x \to 0} \frac{\sin(x)}{x}$$

## Mini project: symbolic analysis of a quadratic function using MATLAB and Python

### Objective

To perform a symbolic analysis of a mathematical function using MATLAB and Python and compare the obtained results.

---

**Problem Statement**

Consider the quadratic function:

$$f(x) = x^2 - 6x + 8$$

---

Perform symbolic analysis of the function using MATLAB and Python.

### Tasks

1. Define the symbolic function $f(x)$.
2. Determine the first derivative $f'(x)$.
3. Find the critical point by solving:

$$f'(x) = 0$$

4. Compute the second derivative $f''(x)$.
5. Classify the critical point using the second derivative test.
6. Evaluate the function at the critical point.
7. Determine the minimum value of the function.
8. Compute the indefinite integral:

$$\int f(x)\, dx$$

9. Generate the Taylor series expansion of $f(x)$.
10. Compare the results obtained using MATLAB and Python and summarize the observations.

CHAPTER 11

# Mini projects and applications

The concepts presented in this book can be used to solve many real-world problems in science, engineering, business, and data analysis. This chapter is a set of mini projects aimed at strengthening problem-solving skills in MATLAB and Python. The projects deal with mathematical modeling, scientific computing, data analysis, visualization, symbolic computation, and numerical techniques. Section A has guided projects with complete solutions and Section B has practice projects that encourage independent execution and investigation.

## Section A

### Project A1: compound interest calculator

#### Objective

To develop a program that computes the future value of an investment using the compound interest formula and analyzes the growth of money over time.

**Problem statement**
A person invests ₹75,000 at an annual interest rate of 7% compounded yearly for 12 years. Determine the final amount accumulated at the end of the investment period.

The compound interest formula is given by:

$$A = P(1 + r)^n$$

where $P$ is the principal amount, $r$ is the annual interest rate, $n$ is the number of years, and $A$ is the final amount.

**MATLAB solution**

```
P = 75000;
r = 0.07;
n = 12;
A = P*(1+r)^n;
interest = A – P;
```

 | HTTPS://DOI.ORG/10.1515/9783112207758-011

```
disp(A)
disp(interest)
```

**Python solution**

```
P = 75000
r = 0.07
n = 12
A = P*(1+r)**n
interest = A – P
print(A)
print(interest)
```

**Output:**

```
168850.64
93850.64
```

**Discussion**

The investment grows from ₹75,000 to approximately ₹168,850.64 after 12 years. The total interest earned during the investment period is approximately ₹93,850.64. This project demonstrates the practical application of mathematical formulas in financial analysis using MATLAB and Python.

## Project A2: image matrix analysis

To perform basic image-processing operations on a grayscale image represented as a matrix and extract useful statistical information from the image data.

**Problem statement**

A grayscale image is represented by the matrix:

$$A = \begin{bmatrix} 100 & 120 & 150 \\ 110 & 140 & 160 \\ 130 & 170 & 200 \end{bmatrix}$$

Analyze the image matrix and determine:

The transpose of the matrix
The maximum intensity value
The minimum intensity value
The average intensity value
The range of intensity values

**MATLAB solution**

```
A = [100 120 150;
     110 140 160;
     130 170 200];

AT = A';

max_intensity = max(A(:));
min_intensity = min(A(:));
avg_intensity = mean(A(:));

range_intensity = max_intensity – min_intensity;

disp(AT)
disp(max_intensity)
disp(min_intensity)
disp(avg_intensity)
disp(range_intensity)
```

**Python solution**

```
import numpy as np

A = np.array([[100,120,150],
              [110,140,160],
              [130,170,200]])

AT = A.T

max_intensity = np.max(A)
min_intensity = np.min(A)
avg_intensity = np.mean(A)

range_intensity = max_intensity – min_intensity

print(AT)

print(max_intensity)
print(min_intensity)
print(avg_intensity)
print(range_intensity)
```

**Output:**
Transpose Matrix

$$\begin{bmatrix} 100 & 110 & 130 \\ 120 & 140 & 170 \\ 150 & 160 & 200 \end{bmatrix}$$

### Statistical results

| Quantity | Value |
|---|---|
| **Maximum intensity** | 200 |
| **Minimum intensity** | 100 |
| **Average intensity** | 142.22 |
| **Intensity range** | 100 |

### Discussion

The matrix represents a grayscale image in which larger values correspond to brighter pixels. The average intensity of approximately 142.22 indicates a moderately bright image. The intensity range of 100 shows a significant variation between the darkest and brightest pixels. Such statistical measures are commonly used in image enhancement, segmentation, and computer vision applications.

## Project A3: population growth analysis

### Objective

To analyze population growth over a period of time and determine annual growth rates and average growth trends using MATLAB and Python.

### Problem statement

A town records its population over five consecutive years as:
[1000, | 1120, | 1260, | 1415, | 1580]
Determine:

1. Annual growth percentages
2. Average growth rate
3. Population increase from the first year to the last year

**MATLAB solution**

```
population = [1000 1120 1260 1415 1580];
growth = [];
for i = 2:length(population)
   rate = ((population(i)-population(i-1)) ...
      / population(i-1))*100;
growth = [growth rate];
end
average_growth = mean(growth);
population_increase = population(end) – population(1);
disp(growth)
disp(average_growth)
disp(population_increase)
```

**Python solution**

```
population = [1000,1120,1260,1415,1580]
growth = []
for i in range(1,len(population)):
   rate = ((population[i]-population[i-1])
      / population[i-1])*100
growth.append(rate)
average_growth = sum(growth)/len(growth)
population_increase = population[–1] – population[0]
print(growth)
print(average_growth)
print(population_increase)
```

**Output:**
**Annual Growth Rates (%)**

| Year Interval | Growth Rate (%) |
|---|---|
| **Year 1 → Year 2** | 12.00 |
| **Year 2 → Year 3** | 12.50 |
| **Year 3 → Year 4** | 12.30 |
| **Year 4 → Year 5** | 11.66 |

**Average growth rate**

$\approx$ **12.12%**

**Total population increase**

$$1,580 - 1,000 = 580$$

**Discussion**
The population is growing at a relatively steady annual rate of about 12%. The population increased by 580 individuals over the 5-year period. These analyses are frequently used in demographic studies, urban planning, resource allocation, and economic forecasting. In this project, we will see how we can use MATLAB and Python to unite arrays, loops, arithmetic operations, and basic statistical calculations to analyze real world data.

## Section B: independent projects

The following projects are for individual practice. Only the objectives, problem statements, and implementation hints are provided. Readers are encouraged to build full-blown solutions in MATLAB and Python throughout this book based on the concepts introduced.

### Project B1: compound interest calculator

#### Objective

To develop a program that computes the future value of an investment using the compound interest formula and analyzes the growth of money over time.

**Problem statement**
A person invests ₹75,000 at an annual interest rate of 7% compounded yearly for 12 years. Determine the final amount accumulated at the end of the investment period.

The compound interest formula is given by:

$$A = P(1 + r)^n$$

where $P$ is the principal amount, $r$ is the annual interest rate, $n$ is the number of years, and

$A$ is the final amount.

**Hints**

- Store the principal amount, interest rate, and investment period in variables.
- Apply the compound interest formula.
- Display the final accumulated amount.
- Calculate the total interest earned.
- Compare the final amount with the initial investment.
- Extend the program to display the investment value year by year.

### Project B2: image matrix analysis

To perform basic image-processing operations on a grayscale image represented as a matrix and extract useful statistical information from the image data.

**Problem statement**

A grayscale image is represented by the matrix:

$$A = \begin{bmatrix} 100 & 120 & 150 \\ 110 & 140 & 160 \\ 130 & 170 & 200 \end{bmatrix}$$

Analyze the image matrix and determine:

1. The transpose of the matrix
2. The maximum intensity value
3. The minimum intensity value
4. The average intensity value
5. The range of intensity values

**Hints**

- Store the image data as a matrix (MATLAB) or NumPy array (Python).
- Compute the transpose of the matrix.
- Determine the maximum and minimum intensity values.
- Calculate the mean intensity of all pixels.
- Find the intensity range using:

$$\text{Range} = \text{maximum intensity} - \text{minimum intensity}$$

- Interpret what the obtained statistics indicate about the brightness distribution of the image.

- Extend the project by visualizing the matrix as a grayscale image using appropriate plotting functions.

### Project B3: Armstrong number detector

#### Objective

To develop a program that determines whether a given number is an Armstrong number by analyzing its individual digits and applying mathematical operations.

**Problem statement**
Write a program that accepts a positive integer and determines whether it is an Armstrong number. Test the program using the number 153.

**Hints**
- Use loops to extract individual digits.
- Compute the cube of each digit.
- Sum the cubes of all digits.
- Compare the result with the original number.
- Display an appropriate message.

### Project B4: Newton's law of cooling

#### Objective

To model the cooling process of a heated object and analyze temperature variation with time.

**Problem statement**
The temperature of an object follows Newton's law of cooling:

$$T = T_s + (T_o - T_s)e^{-kt}$$

where $T_s$ is the surrounding temperature, $T_o$ is the initial temperature, $k$ is the cooling constant, and $t$ is the time.

Develop a program to compute and visualize the cooling behavior over a specified time interval.

**Hints**
- Create a user-defined function.
- Generate temperature values for different time instants.

- Store results in arrays.
- Plot temperature versus time.
- Interpret the cooling curve.

## Project B5: eigenvalue analysis of a matrix

### Objective

To perform matrix analysis by determining eigenvalues and eigenvectors of a square matrix.

**Problem statement**

Consider the matrix:

$$A = \begin{bmatrix} 4 & 1 \\ 2 & 3 \end{bmatrix}$$

Determine the eigenvalues and corresponding eigenvectors.

**Hints**

- Define the matrix.
- Use built-in functions for eigenvalue computation.
- Display eigenvalues and eigenvectors separately.
- Verify the relationship:

$$A\mathbf{v} = \lambda\mathbf{v}$$

for each eigenvalue.

## Project B6: Taylor series approximation

### Objective

To approximate a mathematical function using a finite number of terms from its Taylor series expansion.

**Problem statement**

Generate the Taylor series approximation of $e^x$ using the first five non-zero terms and compare the approximation with the exact function value.

**Hints**
- Generate symbolic or numerical series terms.
  Evaluate the approximation at selected values of $x$.
- Compute the exact function value.
- Determine the approximation error.
- Discuss the effect of increasing the number of terms.

## Project B7: population forecasting

### Objective

To analyze historical population data and predict future population growth.

**Problem statement**
Given the population data for the previous 5 years:

$$[1000, | 1120, | 1260, | 1415, | 1580]$$

Estimate the population for the next 5 years.

**Hints**
- Compute annual growth rates.
- Determine the average growth rate.
- Use the average growth rate for prediction.
- Display projected population values.
- Plot historical and predicted data.

## Project B8: numerical differentiation

### Objective

To estimate the derivative of a function numerically and compare it with the analytical derivative.

**Problem statement**
Consider the function:

$$f(x) = x^3 + 2x^2$$

Determine the derivative at selected points using numerical differentiation techniques.

**Hints**

- Use finite-difference formulas.
- Compute the analytical derivative.
- Compare numerical and exact values.
- Calculate approximation errors.
- Discuss the effect of step size.

### Project B9: symbolic equation solver

#### Objective

To solve algebraic equations symbolically and verify the obtained solutions.

**Problem statement**

Solve the quadratic equation:

$$x^2 - 7x + 12 = 0$$

using symbolic computation techniques.

**Hints**

- Define symbolic variables.
- Use symbolic equation-solving functions.
- Obtain exact roots.
- Substitute roots into the original equation.
- Verify the correctness of the solutions.

### Project B10: student performance analyzer

#### Objective

To analyze student academic performance using statistical and graphical techniques.

**Problem statement**

Consider the marks obtained by students in a class:

$$[78, \mid 85, \mid 92, \mid 67, \mid 88, \mid 95, \mid 74, \mid 81]$$

Develop a program to analyze the performance of the class.

**Hints**
- Compute mean, median, and standard deviation
- Determine highest and lowest marks
- Classify students into performance categories
- Generate suitable charts for visualization
- Interpret the obtained results

### Project B11: experimental data curve fitting

#### Objective

To develop a mathematical model that best represents a set of experimental observations.

**Problem statement**
Given a set of experimental data points, determine a suitable polynomial model and use it to predict future values.

**Hints**
- Store the data points in arrays.
- Perform polynomial fitting.
- Plot original and fitted curves.
- Evaluate prediction accuracy.
- Estimate values beyond the measured range.

### Project B12: symbolic optimization of a function

#### Objective

To determine the critical points and optimal values of a mathematical function using symbolic techniques.

**Problem statement**
Consider:

$$f(x) = x^3 - 3x^2 - 9x + 5$$

Determine the critical points and identify maxima and minima.

**Hints**

- Compute the first derivative.
  Solve $f'(x) = 0$.
- Compute the second derivative.
- Apply the second derivative test.
- Evaluate the function at critical points.

### Project B13: differential equation solver

#### Objective

To obtain the symbolic solution of a differential equation and analyze its behavior.

**Problem statement**

Solve the differential equation:

$$\frac{dy}{dx} + y = e^{-x}$$

using symbolic computation techniques.

**Hints**

- Define symbolic functions.
- Formulate the differential equation.
- Use symbolic differential equation solvers.
- Display the analytical solution.
- Verify the solution by substitution.

## Chapter summary

In this chapter, a sequence of project-based activities was used to demonstrate the practical application of MATLAB and Python. The guided projects demonstrated how to create complete computational solutions, while the practice projects encouraged independent problem-solving. The projects covered numerical computation, data analysis, matrix operations, visualization, optimization, and symbolic mathematics. These activities form the basis of the use of computational tools for solving engineering, scientific, and analytical problems in the real world.

# Authors

*Dr Mamta Kapoor*
is a mathematician and researcher whose research interests include numerical approximation of linear and nonlinear partial differential equations, semianalytical solutions of fractional partial differential equations, fractional calculus, fuzzy calculus, scientific computing, numerical analysis, and fluid mechanics. She is currently working as associate professor for research in the Department of Mathematics, Faculty of Engineering and Technology, Marwadi University Research Center, Marwadi University, Rajkot, Gujarat, India. She has several publications in reputed national and international journals indexed in Scopus. In addition to her significant research contributions, she is the author of *Practical MATLAB and Python: A Comparative Approach to Problem-Solving to Code Like a Pro*, a work that exemplifies her dedication to integrating theoretical mathematics with computational and applied problem-solving methodologies. Moreover, she has authored the books entitled *The Data Science Interview Series – Volume 1: Python, Data Analysis & ML Foundations* and *The Data Science Interview Series – Volume 2: Advanced ML, Deep Learning & Practice* and *Practical Pandas: Data Wrangling, Analysis, and Visualization*. Her scholarly efforts consistently emphasize the practical implementation of mathematical concepts in modern engineering and scientific contexts. She has also pursued several certifications in data science and related fields, demonstrating her keen interest in interdisciplinary applications of mathematics and modern computational techniques.

*Dr Geeta Arora*
received her PhD in mathematics from the Indian Institute of Technology (IIT) Roorkee, India, in 2011. She is currently serving as a professor in the Department of Mathematics at Lovely Professional University, Punjab, India. With over 12 years of teaching and research experience, she has published numerous research articles in reputed national and international journals indexed in Scopus and other leading databases.

She has contributed several book chapters to national and international publications. In addition, she has edited three international books on numerical methods published by Taylor & Francis, Elsevier, De Gruyter, Wiley, and IGI Global. She is currently involved in editing and authoring several books with renowned publishers, including Apple Academic Press (AAP), Taylor & Francis, De Gruyter, Wiley-Scrivener, and River Publishers.

Her primary research interests include numerical methods, scientific computing, and statistics. She has organized and conducted numerous workshops on MATLAB and Vedic Mathematics for students, researchers, and faculty members both within and outside the university. She has successfully supervised nine PhD scholars and is currently guiding six doctoral researchers.

 | HTTPS://DOI.ORG/10.1515/9783112207758-012

# Index



 | HTTPS://DOI.ORG/10.1515/9783112207758-013